Instrumental Analysis
for
Water Pollution Control

Copyright© 1971 by Ann Arbor Science Publishers, Inc.
P. O. Box 1425, Ann Arbor, Michigan 48106

Library of Congress Catalog Card No. 70-141230
ISBN 250-97503-3

1st Printing August, 1971
2nd Printing January, 1972
3rd Printing January, 1973

Instrumental Analysis

for

Water Pollution Control

Khalil H. Mancy
Editor
Professor of Environmental Chemistry
School of Public Health
The University of Michigan

This text has been prepared for the
World Health Organization, Regional
Office for Europe, for use in the
United Nations Development Programmes
SF/WHO-assisted programme "Protection
of River Waters Against Pollution,
Poland 0026"

 **ann arbor science publishers inc.**

Contributors

H. E. Allen
Lecturer in Environmental Health
School of Public Health
The University of Michigan
Ann Arbor, Michigan

J. B. Andelman
Associate Professor of Water Chemistry
Graduate School of Public Health
The University of Pittsburgh
Pittsburgh, Pennsylvania

R. F. Christman
Associate Professor of Applied Chemistry
Department of Civil Engineering
The University of Washington
Seattle, Washington

Rolf Deininger
Associate Professor of Environmental Health
School of Public Health
The University of Michigan
Ann Arbor, Michigan

K. H. Mancy
Professor of Environmental Chemistry
School of Public Health
The University of Michigan
Ann Arbor, Michigan

F. C. Polcyn
Research Engineer
Willow Run Laboratories
Infrared and Optics Laboratory
The University of Michigan
Ann Arbor, Michigan

C. T. Wezernak
Research Engineer
Willow Run Laboratories
Infrared and Optics
The University of Michigan
Ann Arbor, Michigan

Efforts to protect the quality of our environment are highly dependent on our ability for measurement of pollution. *Subjective* response to environmental quality deterioration in terms of perceived awareness (or annoyance to such things as taste, odor or noise are ultimately of great significance. Nevertheless, little headway can be made in setting standards, devising quality indices, and guarding against pollution of natural resources unless there is a basis in *objective* physical measurement. This implies the ability to measure reliably and quantitatively pertinent environmental quality parameters, establish baseline values, and monitor changes in these parameters both in location and time.

Presently, our efforts to control water pollution are limited by a lack of adequate measurement capabilities. Large amounts of funds are being appropriated for the removal of pollutants that cannot be measured adequately or cannot be measured at all. This is particularly true in the case of the analysis for trace metals, carcinogens, and persistent organics. In addition, there is a pressing need to *reduce to practice some of the more sophisticated analytical procedures and demonstrate their applicability to water pollution control programs.*

Water quality characterization relies heavily on the use of instrumental and automated measurement techniques. Gone is the time when water pollution measurements consisted only of no more than simple titrimetric and gravimetric procedures. Nowadays, there is an ever broadening need for more accurate and precise measurement of water quality which can only be provided by advanced instrumental techniques. Nevertheless water pollution measurement cannot be done solely by laboratory instruments, regardless of how sophisticated they may be. Water quality surveillance programs rely heavily on field measurement, *in situ* analysis and remote sensing techniques. *These constitute the new challenges in instrumental analysis for water pollution control.*

This book is addressed to scientists and engineers concerned with water pollution characterization and control. Emphasizing the application of instrumentation, including automatic and remote measurement systems for water quality characterization, the book is presented with the premise that the reader does not have to be a chemist by background. This book can be used by practitioners in the field of water pollution control who are faced with the day-to-day task of using instrumental and automated measurements or the interpretation of these measurements. In addition, the test is designed to meet the needs of graduate or undergraduate programs of aquatic sciences and engineering, *e.g.*, sanitary engineering, environmental health and aquatic or oceanographic science programs.

The authors wish to acknowledge the support of the World Health Organization for the preparation of this text for use in water pollution control training programs. Special thanks are due to Mr. George Ponghis, Mr. Alex Gilad and Dr. Michael Sues, of the Environmental Health Division, World Health Organization, Regional Office for Europe, for their encouragement and support of this effort.

Khalil H. Mancy
Ann Arbor, Michigan
August, 1971

Contents

Chapter I. Introduction   1
        by K. H. Mancy

    Chemical Analysis for Water Pollution Control . .   1
    What Is Unique About Water Analysis?   2
    What Are the Responsibilities of the Water Analyst? 3
    Instrumental and Automated Analyses   3

Chapter II. Design of Measurement Systems   7
        by K. H. Mancy

    Objectives of Analysis   7
    Parameters for Analysis   11
    Methods of Analysis   16
    Intensive Versus Extensive Measurement   18
    Primary and Secondary System Characteristics . .   21
        Primary sensor characteristics   21
        Secondary sensor characteristics   24
    References   24

Chapter III. Design of Sampling Programs   27
        by H. E. Allen

    Sampling Programs   27
    Sample Collection   29
    Sample Preservation   30
    Corollary Field Data   31
    References   31

Chapter IV. Concentration and Separation Techniques   33
          by J.B. Andelman and R.F. Christman

Introduction 33
Carbon Adsorption 35
Freeze Concentration 37
Liquid-Liquid Extraction 38
    Principles 39
    Applications 41
Ion Exchange 43
    Ion Exchange Equilibria and Kinetics 43
    Column Operation 45
    Applications 46
Chromatography 50
    General principles 50
    Gas-Liquid chromatography 52
        Separation Process 52
        Columns 52
        Sample Introduction 55
        Carrier Gas 55
        Detectors 55
            Thermal conductivity detection 56
            Flame ionization detection 56
            Electron capture detection 57
            Microcoulometric detection 57
        Data Interpretation 57
    Thin-Layer Chromatography 58
Other Techniques 59
References 60

Chapter V. Elements of Instrumental Analysis   . . .   63
          by K.H. Mancy

Chemical Spectrophotometry - General 63
Molecular Absorption Spectrophotometry 65
Molecular Fluorescence Spectrophotometry 70
Atomic Absorption Spectrophotometry 75
Atomic Fluorescence Spectrophotometry 77
Infrared Absorption Spectrophotometry 80
Ultraviolet Absorption Spectrophotometry 88
Electrochemical Analysis 91
    Methods based on passage of faradaic current . 92
Measurement of Current-Potential Curves 101
    Methods based on electrodes at equilibrium . . 105
    Methods in which faradaic current
    is unimportant 107
References 110
Selected Supplementary References 113

Chapter VI.   Continuous Monitoring Systems 115
          by H.E. Allen

Systems Not Requiring Reagent Additions 115
Systems Requiring Addition of Reagents 117
Data Collection 118
Sensors 121
Analysis and Control of Biological Processes . . 127
References 132

Chapter VII.   Automated Chemical Analyses 135
          by H.E. Allen

System Design 135
Automation of Manual Procedures 145
Response Characteristics of Flow Systems 150
Analytical Methods 154
    Ammonia 154
    Chloride 155
    Chemical oxygen demand (COD) 156
    Iron 156
    Nitrite 157
    Nitrate 157
    Kjeldahl nitrogen 158
    Manganese 158
    Phosphate 158
    Silica 160
    Other analyses 160
References 160

Chapter VIII.   Remote Optical Sensing Techniques . . 165
          by C.T. Wezernak and F.C. Polcyn

Introduction 165
General Considerations 167
Scattering 167
Fluorescence 171
Infrared 174
    Instrumentation 176
        Multispectral sensor system 177
    Fraunhofer line discriminator 179
        Barringer correlation spectrometer 180
Selected Examples of Data Acquisition Using the
Multispectral Scanner 181
    Temperature Analysis 182
    Effluent Outfalls 187
    Oil Pollution 187
    Industrial Effluents 191
Conclusions 200
References 202

Chapter IX. Physical Characteristics 205
        by K.H. Mancy

Density and Viscosity 205
Temperature 206
Electrical Conductance 208
    Electrical conductivity 208
Turbidity 210
References 212

Chapter X. Analysis for Organic Pollutants 215
        by R.F. Christman

Nonspecific Analysis 215
Separation and Identification of
Organic Pollutants 218
    Application of GLC to pesticide analysis . . . 218
    Application of GLC to phenol analysis 219
    Application of GLC to volatile acid and
    digester gas analysis 222
    Application of TLC to phenol analysis 222
        Plate preparation 223
        Application of sample 223
        Development 223
            Developing chamber 223
            Solvent systems 224
        Detection 224
            Drying 224
            Fluorescence 224
            Chromogenic reagents 224
Ultraviolet and Infrared Identification of
Organic Pollutants 224
    Identification of pesticides 226
    Identification of phenols 226
References 228

Chapter XI. Analysis for Metal Pollutants 231
        by J.B. Andelman

Physicochemical Characteristics of Metal Ions
in Aquatic Environments 231
Molecular Absorption Photometry and Colorimetry . 234
Atomic Absorption Spectrophotometry 237
Other Flame Techniques 239
Emission Spectroscopy 240
Neutron Activation Analysis 241
Ion Selective Electrodes 241
Polarography 243
Anodic Stripping Voltammetry 244
References 248

Chapter XII. Analysis for Inorganic Anions 251
            by K.H. Mancy

    Separation Techniques 251
    Spectrophotometric Techniques 251
    Electrochemical Techniques 252
Nitrogen Compounds 255
Phosphorus Compounds 258
References 258

Chapter XIII. Dissolved Gases 261
            by K.H. Mancy

Separation Techniques 261
Dissolved Oxygen 262
    Physicochemical properties of dissolved oxygen 262
    Instrumentation modifications of Winkler test 264
    Coulometric titration 265
    Gas exchange methods 267
        Gas exchange unit 267
        Oxygen-detecting units 269
    Gas chromatographic methods 273
    Radiometric method 275
    Conductometric methods 276
    Voltammetric methods 278
    Voltammetric membrane electrode 279
        Sensitivity 285
        Temperature coefficient 286
        Application 289
References 294

Chapter XIV. Analysis of Data 299
            by R. Deininger

Curve Fitting 300
    Example 1 301
    Example 2 305
    Example 3 306
Use of Probability Paper 309
    Example 4 310
    Example 5 313
Tests for Statistical Significance 315
Reliability of Statistical Results 318
References 318

Chemical Analysis for Water Pollution Control

The seriousness of the problem of water pollution and its impact on environmental health is presently the focus of international attention. Never was there a time in the history of man when he was so deeply concerned with the quality of his water. Much legislation has been passed and appropriations have been made to restore the quality of man's environment. The chemical industry, accused of being a main contributor to pollution, is undertaking expensive programs designed to limit emmission and improve disposal of industrial wastes as well as to ensure fitness and appropriateness in the use of industrial products.

The analytical chemist, as a member of a team combining various disciplines such as other scientists, engineers, lawyers, sociologists, physicians, and economists, plays a significant role in the total effort of water pollution control. His role requires him to characterize the nature and magnitude of the problem, assist in executing a control measure or treatment process, and last but not least, provide a continuous surveillance and monitoring of the final product, clean water. To this effect the analytical chemist is the "eyes" and "ears" of any control measure, for without valid, meaningful analysis our efforts to control pollution will be fruitless.

There was a time when all that water analysis did consisted of simple titrimetric and gravimetric procedures. Nowadays, with the obvious increase in the magnitude and the ways by which man is polluting his waters, there is an ever broadening need for more accurate and precise

information on water quality. The determination of a
specific element or compound in a water sample frequently
requires extensive separation work to remove interferring
constituents before identification and quantitative
measurements can be performed. All too frequently, this
is followed by one or multiple identification procedures
which may involve the use of such sophisticated equipment
as emission and absorption spectrometry, mass spectroscopy, NMR, X-ray, thermochemical and eletrochemical
techniques. For example, the analysis for certain organic
compounds in river waters may first require concentration
by adsorption on carbon columns, or partial freezing or
reverse osmosis membrane techniques. This may be followed by separation with liquid-liquid extraction, gas-
liquid chromatography, or thin-layer chromatographic
techniques and, finally, the identification and measurement may be done by infrared spectrophotometry.

What is Unique About Water Analysis?

The chemical analysis of natural and waste waters is
one of the most challenging tasks that is likely to con-
front any analytical chemist. The test solution is, in
most cases, a complicated heterogeneous system that rarely
lends itself to simple analytical techniques. "Water"
analysis requires not only the subtle correlation of
theory and experience of analytical principles but also
a keen insight into the nature of interferences and other
problems associated with the methodology. Interferences
may be quite unique to a particular water system and in
many ways yield inaccurate data and cause misleading conclusions. Water analysis requires the ability to properly
interpret analytical results in correlation to pertinent
field observation and the history of the water.

Water pollution control can be looked at as a specialized technology and, like any other industry, it has its
own specific analytical needs. Analysis may be required
to (a) determine the compliance of certain waters or
waste waters with quality standards for an intended use,
production control or disposal in a receiving stream,
(b) estimate possible detrimental effects of waste effluents on the quality of a receiving water for subsequent
downstream use, and (c) evaluate treatment requirements
in view of water use.

Some of the more unique analytical parameters characteristic of the water pollution control industry are BOD
(biochemical oxygen demand), COD (chemical oxygen demand),
taste, odor, color, chlorine demand, hardness, alkalinity,

bioassay using fish or crustaceae and biodegradability tests. The analytical chemist, experienced in water quality characterization, usually makes decisions based on practical intuition as to what parameters and tests to use for a given purpose.

What are the Responsibilities of the Water Analyst?

In the water pollution control field the analytical chemist is expected not just to prescribe a workable procedure but to optimize the technique in order to reduce the time and effort required to obtain the needed information. Besides his responsibility for chemical analysis, a good part of his efforts is devoted to the diagnosis of problems, definition of analytical needs, design of measurement systems and development of a master plan of how to report, evaluate and integrate information with that of other efforts on the same problem.

Unlike other chemists, the water analyst has to be able to work with other professional groups such as engineers, physicians, aquatic biologists, lawyers, etc., and nonprofessional citizen groups on the local, state, and federal levels. He is frequently faced with many questions that may have nothing to do with analysis, and he must make decisions and recommendations on the spot. Based on his ability and initiative, the water analyst can find his job to be most challenging as well as rewarding.

Instrumental and Automated Analyses

Water pollution control programs rely to a great extent on a variety of instrumental and automated analytical techniques. It is not unusual to find water treatment plants equipped with modern laboratories utilizing some of the most advanced instrumentation techniques. Some of these instrumental analysis procedures are highly specialized and require highly qualified personnel.

Perhaps the most challenging and most meaningful type of instrumental analysis for water quality characterization is that which involves *in situ* measurements. This technique is preferred over analytical procedures which remove the water from its natural environment, either in the form of "grab" samples or a continuous flow stream, for subsequent analysis in the laboratory or field station. The main disadvantage of the latter procedure arises with the problem of collecting samples. In most cases, this cannot be done in such fashion as to give a true representation of the test solution. In addition the analyses are usually performed under environmental

3

conditions quite different from those which existed at the sampling site. For example, changes in pressure and temperature may result in the escape of gases which in turn cause chemical or biological transformations of the species under test. In monitoring water quality, the grab sample method not only lacks required statistical accuracy but also can be relatively expensive on a cost per sample basis.

Before the instrumentation era, a classical method of the *in situ* analysis for toxic compounds in water was based on the survival of mosquito fish. In a fashion similar to the 19th century practice of using canaries to detect the presence of toxic gases in coal mines, "Gumbosia affinis" was used to monitor industrial waste effluents. Although this method lacks specificity, bioassay techniques, using various biological indicators, are inherent methods of water quality characterization.

Certain parameters easily lend themselves to *in situ* measurements such as temperature, turbidity, pH, and salinity. Electrochemical techniques, in general, are suited for *in situ* measurements. The effect of interferences, due to the presence of electroactive as well as surface active species in the test solution, can be minimized by covering the indicator electrode with permselective membranes. Examples of such systems are the voltammetric membrane electrode systems used for the *in situ* measurements for dissolved oxygen or certain metal ions. Potentiometric membrane electrode systems using glass or crystal membranes also have been used for measurements of certain cations and ions.

Automated analyses are being used rather extensively to monitor water quality in rivers, streams, and municipal and industrial plant effluents. This is usually done by an on-stream analysis monitoring system utilizing a number of electrochemical sensors for the continuous measurement of temperature, pH, turbidity, dissolved oxygen, chlorides, hardness, redox potential, residual chlorine, and specific conductance. Such monitoring systems are fully automated and should require minimum attendance and service.

Another type of automated analysis, frequently used in water analysis laboratories, is based on automating wet analytical procedures, e.g., Technicon Autoanalyzer. On-stream or repetitive measurements of natural and waste waters can be done with a high degree of precision using such systems. This type of measurement is particularly useful whenever large volumes of water samples are concerned.

4

With the increasing dependency on instrumental analysis
and automatic monitoring systems, there is a rising demand
for highly qualified water analysts. One of the main
problems in water pollution control programs is the lack
of aquatic scientists and engineers trained for the application of advanced instrumentation for water pollution
control characterization. Attempts to remedy the situation include the offering of short training courses by a
number of national and international agencies, such as the
U. S. Public Health Service and the European Environmental
Health Program of the World Health Organization. Also,
several university undergraduate and graduate programs in
Sanitary Engineering and Environmental Health are offering
courses on the application of instrumental analysis for
water pollution control. It is important to stress here
the difference between these training courses and classical courses in instrumental analysis, commonly offered in
departments of chemistry. The main difference lies in the
fact that the former type of training is mission-oriented
and is designed to demonstrate the applications and limitations of instrumental and automated analysis for water
pollution characterization.

Chapter II

Design of
Measurement Systems

There are three basic steps in the design of a measurement system for water quality characterization. These are based on the answers to the following questions:
- (a) Why are the measurements needed? or *"Objectives of Analysis"*
- (b) What parameters should be sought? or *"Parameters of Analysis"*
- (c) How should the measurement be made? or *"Method of Analysis"*

Objectives of Analysis

Definition of the purpose and objectives of analysis is the first step in the design of any measurement system; this includes the definition of particular problems to which solutions are sought. Some of the more common objectives of industrial waste water analysis are as follows:
- (a) Determine the suitability of a water for its intended use and establish the degree of treatment necessary prior to its use.
- (b) Estimate the possible detrimental effects of a waste effluent on the quality of the receiving water for subsequent downstream use.
- (c) Evaluate necessary treatment requirements in view of water reuse.
- (d) Determine the quantities of valuable by-products which could be recovered from a waste effluent.
- (e) Evaluate and optimize industrial processes on a continuous or batch basis.

(f) Provide background information on the present quality of streams and lakes which can be used to demonstrate future changes in their quality.

Natural bodies of fresh water can be classified according to intended use--public water supply, fish or shellfish propagation, recreation, agricultural use, industrial water supply, hydroelectric power, navigation, and disposal of sewage and industrial wastes. Under the terms of the 1965 *Water Quality Act* in the U.S.A., it is required that states adopt water quality criteria for their interstate and coastal waters and a plan implementing and enforcing the criteria adopted.

The water quality criteria and standards relate to specific desired uses for the waters and therefore are based on specific chemical requirements necessary for these uses. The parameters important for these uses are discussed in the section "Selection of Parameters to be Measured."

Water quality criteria may be based on the quality to be maintained in a receiving water, or they may refer to the quality of the effluent itself. Criteria relating to the receiving water are termed "stream standards" while those relating to the waste discharge are termed "effluent standards." Both types of standards have their strengths and weaknesses.

Stream standards, which have been developed to comply with the provisions of the *Water Quality Act*, permit the full capacity of a stream to be utilized in the assimilation of wastes. When stream standards are established, the quality of the receiving stream is maintained at a level above that required for the desired usage. This permits full utilization of the stream's capacity to assimilate wastes and requires that materials discharged into the different streams be subjected to different degrees of treatment prior to discharge. In contrast, if effluent standards were used, all discharges of the same type to streams designated for the same use would be required to receive identical levels of treatment. The level of treatment would be consistent with the capability of advanced technology. The effect of the effluent on the receiving waters would differ and there would be no assurance that the concentration of pollutant in the stream would not exceed that critical to the desired use.

It is readily apparent that either system of standards has its flaws. It is likely that future standards will incorporate features of both stream and effluent standards. At present such standards are coming into effect for waste discharge of phosphate. In the Lake Erie Basin

8

a goal of 90 per cent phosphate removal from industrial and municipal discharges has been suggested.[4] The 90 per cent reduction constitutes an effluent standard whose value was based upon desired stream standards and the capabilities of waste treatment technology.

Chemical analyses associated with the control and treatment of industrial and municipal wastewaters discharged to streams, lakes or oceans depend on the type of water quality standards to which they must conform. The analyses and data interpretation required to ensure that discharge of an effluent will not violate stream standards are usually more complicated than those to ensure compliance with effluent standards. In the former case the analysis is directed not only to characterization of the quality of the waste effluent, but also to determination of its effect on the aquatic ecosystem of the receiving water, while in the latter case only the quality of the waste is of concern.

One common method for estimating the deleterious effects a waste effluent will have on the quality of a receiving water is to treat the wastewater as a complete entity, thus avoiding analyses for particular constituents. For this type of evaluation the wastewater is first diluted to a level corresponding to that occurring in the receiving water. Then certain gross parameters, such as taste, odor, color, and toxicity to fish, are measured depending on the intended water use. While this procedure may give a preliminary indication of the ability of the waste to be assimilated harmlessly in the receiving water, its effectiveness for providing sufficient basis for any significant conclusion is highly doubtful. Among other things, the rate of self-purification of the receiving water is not accounted for in such a test procedure. Self-purification of streams and other receiving waters is a dynamic process involving a multitude of physical, chemical and biological reactions. The rates of biochemical transformation of pollutants are often much more significant than the ultimate assimilative capacity of the stream *per se*. For example, in evaluating the biochemical oxidation demand (BOD) of a particular waste water, determination of the rate constant is at least as important as determination of the five-day BOD. This method of analysis will be discussed in detail in a later section of the chapter.

The analysis of wastewaters which are to be discharged to municipal sewers is done principally to evaluate compliance with certain effluent criteria set by the municipality. Effluent standards in this case are established to protect municipal waste treatment plants from operational interference which might be caused by industrial waste discharged and to protect the sewer structure from damage. Both the municipality and the industry may periodically analyze the waste effluent for purposes of control and assessment of charges, which are usually related to the strength and volume of a particular waste. It is important to point out that waste water which is discharged to municipal sewers usually becomes the responsibility of the municipality. The general requirements for acceptable wastewater for joint treatment with municipal wastes have been discussed in some detail by Byrd and others.

As far as in-plant operations are concerned chemical analysis of industrial waste waters is performed for one or more of the following purposes:

(a) estimation of material balances for processes to permit evaluation of unit efficiencies and to relate material losses to production operations;

(b) evaluation of continuing conformance to limits set for performance efficiency of certain unit processes;

(c) evaluation of the effectiveness of in-plant processes, modifications and other measures taken for reduction of losses;

(d) determination of sources and temporal distributions of waste loads for purposes of by-product recovery or segregation of flows, relative to strength and type, for separate treatment;

(e) provision for immediate recognition of malfunctions, accidents, spills or other process disturbances;

(f) determination of the type and degree of treatment required for recovery of certain substances from waste effluents;

(g) evaluation of conformance to standards set for effluent quality and/or stream quality;

(h) provision for control of treatment and discharge of waste effluents according to present standards and/or according to variations in the conditions of the receiving water; and

(i) provision of a current record of costs associated with discharge of waste effluents to municipal sewers when such costs are at least partially based on the chemical characteristics of the waste.

Some of the objectives of industrial waste water analysis listed above are of course exploratory in nature and are therefore performed only occasionally, while others are related to continuous or regular monitoring and control.

An important objective of many surveys of lakes and streams is to provide a base of information for evaluating future changes in water quality. Such data have been used by Beeton[3] and others in demonstrating the eutrophication of the Great Lakes. These data are important sources for other investigators. The U. S. Geological Survey has compiled data on the quality and quantity of water of surface waters of the United States[2] and has publications on the quality of surface water for irrigation[7] and of municipal water supply for industrial use.[6] Other water quality data have been published by the Public Health Service and Federal Water Pollution Control Administration.[1,5]

Parameters for Analysis

After defining the objectives of analysis, the next step in the design of measurement systems is to decide on particular constituents for which analyses are to be made and what methods are to be employed. The analyst experienced in water quality characterization can often make the proper decision based on practiced intuition. In most cases, however, certain rather well-defined guidelines should be followed.

Depending on the intended subsequent use of a receiving water, the parameters listed in Table 1 are of significance for water quality characterization. These should serve as guidelines for analysis of wastewater quality for purposes of treatment and control.

The choice of parameters for analysis depends primarily on the type of information sought. Certain tests are frequently used for the identification of various types of pollution associated with industrial wastewaters. For example, Table 2 lists a number of tests and their significance.

Some of the most important and most frequently used tests in the analysis of water are the nonspecific tests listed in Table 3. These tests often measure a property of a group of substances. For example alkalinity indicates the capacity of the water to neutralize hydrogen ions. Additionally a physical parameter such as density or a physiological property such as odor may be used. Many of these tests are used to determine the suitability of natural waters for industrial or municipal use and to determine the type and degree of treatment needed to render them acceptable.

11

Table 1

Parameters for Water Quality Characterization

Water Use	Quality Parameters
Domestic Water Supply	1. Color, odor, taste. 2. Organic content: chlorine demand, COD, BOD, TOC, phenols. 3. Carcinogens and toxic compounds, insecticides, pesticides, detergents. 4. Turbidity, salinity. 5. Alkalinity, pH. 6. Total hardness, Ca, Mg, Fe, Si, etc. 7. Pathogenic organisms, total bacterial count (37°C), *E. coli* count, plankton count.
Fish, shellfish, wildlife and recreation	1. Color, odor. 2. Toxic compounds. 3. Turbidity, floating matter, sludge deposits, salinity. 4. Temperature. 5. Dissolved oxygen, BOD. 6. Alkalinity, pH.

(cont)

Table 1
(cont)

Water Use	Quality Parameters
	7. Pathogenic organisms, plankton count.
	8. Nitrogen, phosphorous, etc. (inorganic nutrients which support algae blooms and other undesirable aquatic growth.
Agricultural irrigation	1. Salinity and Na—Ca content.
	2. Alkalinity, pH.
	3. Pesticides, growth regulators, etc.
	4. Persistent synthetic chemicals (*e.g.*, polyethylene derivatives, asphalt sprays, etc.).
	5. Pathogenic organisms.
Watering of livestock	1. Salinity.
	2. Toxic compounds.
	3. Pathogenic organisms.
	4. Plankton count.

Table 2

Significance of Parametric Measurements

Test or Determination	Significance
Dissolved solids	Soluble salts may affect aquatic life or future use of water for domestic or agricultural purposes.
Ammonia, nitrites, nitrates, and total organic nitrogen	Degree of stabilization (oxidation) or organic nitrogenous matter.
Metals	Toxic pollution.
Cyanide	Toxic pollution.
Phenols	Toxic pollution, odor, taste.
Sulfides	Toxic pollution, odor.
Sulfates	May affect corrosion of concrete; possible biochemical reduction to sulfides.
Calcium and magnesium	Hardness.
Synthetic detergents	Froth, toxic pollution.

Table 3

Nonspecific Water Quality Parameters

Physical Parameters	Chemical Parameters	Physiological Parameters
Filterable residues	Hardness	Taste
Salinity	Alkalinity and acidity	Odor
Density	Biochemical oxygen demand (BOD)	Color
Electrical conductance	Chemical oxygen demand (COD)	Suspended matter
	Total carbon	Turbidity
	Chlorine demand	

Some of the more frequently measured parameters in pollution studies are listed in Table 4. The specific chemical analyses to be performed in a water pollution study will depend on the types of materials discharged and on the desired uses of the receiving water.

Table 4

Tests used for the Measurement of Pollution of Natural Waters

Nutrient Demand	Specific Nutrients	Nuisances	Toxicity
Dissolved oxygen	Nitrogen:	Sulfide	Cyanide
Biochemical oxygen demand	Ammonia	Sulfite	Heavy metals
Chemical oxygen demand	Nitrate	Grease and oil	Pesticides
Total carbon	Nitrite	Detergents	
	Organic nitrogen	Phenols	
	Phosphorus:		
	Orthophosphate		
	Polyphosphate		
	Organic phosphorus		

Methods of Analysis

Following the establishment of the objectives of the measurement program and the selection of parameters for analysis, suitable analytical methods are then selected. There are no prescribed procedures which are applicable to all situations but the best method for any given

16

situation must be based upon consideration of many factors. Some of the more important factors are:
  (a) required sensitivity,
  (b) accuracy of method,
  (c) presence of interferences,
  (d) number of samples to be analyzed,
  (e) necessity of field or *in situ* analyses,
  (f) speed required for results,
  (g) availability of required instruments,
  (h) number and skill of laboratory personnel, and
  (i) required use of standard or referee methods.
  When selecting an analytical method the analyst should
carefully distinguish between specific and nonspecific
methods of analysis and between methods which measure intensive properties and those which measure extensive
properties. A number of nonspecific tests used in water
quality characterization are listed in Table 3 and have
been discussed earlier in this chapter. When analyzing
organics, the analyst must often choose between specific
and nonspecific methods. For example, waste effluents
may be analyzed for total carbon (nonspecific) or for
specific organic compounds. Even if the analyst chooses
the specific analysis, he must distinguish between methods
which measure a class of compounds (*e.g.*, the aminoanti-
pyrine method for phenols) and those for specific compounds such as gas chromatographic analysis for individual
phenols.
  Another point to be considered when selecting analytical methods concerns the collection, transportation and
storage of samples. Screening tests should be conducted
for purposes of approximating required sample volumes,
establishing desirable sites for and frequency of sampling, and providing a rough estimate of the waste composition and strength.
  Listings of "standard" and "recommended" methods for
analysis of natural waters and wastewaters are to be
found in a variety of publications sponsored by several
water works, pollution control, and public health agencies
and organizations in this country and abroad. In addition, in several instances certain private industries have
found it desirable to formalize listings of more specific
methods for analysis of particular types of industrial
wastewater.
  While general procedures of analysis for specific waste
constituents are highly useful, the industrial-wastewater
analyst must be careful to guard against over-reliance
upon such procedures, and against the possibility of
being lulled into a false sense of security by results

17

obtained from application of such procedures in instances
where they may not be applicable. Indiscriminate application of general-purpose methods for analysis without due
consideration of specific interferences and other problems
must be avoided. Standardization of procedures should
be made only after these procedures have been thoroughly
evaluated in terms of particular analytical requirements.
Continuing use of such standard methods without modification should then be subject to the condition that the
characteristics of the waste being analyzed do not change
significantly over the duration of the analytical program.
Just as the skilled medical doctor will not prescribe
treatment or medication until he has carefully examined
the patient *in toto*, so the analytical chemist should
select his approach to the analysis of an industrial
wastewater only after making a careful diagnosis of the
total problem. This diagnosis should include consideration of: (a) objectives of the analysis; (b) requirements
of speed, frequency, accuracy, and precision of analysis;
(c) effects of interferences; and (d) effects of system-
atic and environmental conditions on sampling and
measurements.

Intensive Versus Extensive Measurement

Physical and chemical characterization of water quality
can be categorized conveniently as intensive versus extensive measurements. This categorization should not be
considered in terms of rigorous thermodynamic entities but
rather in terms of conceptual quantitative properties of
the system under investigation. It is for reasons of convenience, which will become apparent later, that we should
differentiate between intensive and extensive water quality
parameters.

In textbook terminology extensive properties are additive in the sense that the total value of a system is the
sum of the individual values for each of its constituent
parts. Conversely, intensive properties are not additive,
and can be specified for any system without reference to
the size of that system. This is illustrated in Table 5.

The chemical potential or the molar free energy change

$(\frac{\partial G}{\partial n})_{T,P}$ is further defined:

$$\mu - \mu^o = RT \ln a \qquad (1)$$

where $\underline{a}$ is the activity given in a molar scale and μ^o is
the chemical potential at a reference state. The activity

Table 5

Extensive and Intensive Parameters

System Designation	Intensive Parameter	Extensive Parameter	Work Done by the System
Gravitational	Height (h)	Mass (m)	$= mg(h_2-h_1)$
Thermal	Temperature (T)	Heat capacity (Cp)	$= Cp(T_2-T_1)$
Electrical	Voltage (E)	Charge (q)	$= q(E_2-E_1)$
Chemical	Chemical potential (μ)	Number of moles (n)	$= n(\mu_2-\mu_1)$

where h = height, cm
        m = mass, g
        g = acceleration, cm/sec^2
        E = voltage, volt
        q = charge, coulomb
        T = temperature, degree centigrade
      Cp = heat capacity, cal/deg/mole
        μ = the chemical potential, cal/mole
        n = number of moles

can be related empirically to concentration C by the equation:

$$a = \gamma C \tag{2}$$

where γ is the activity coefficient. Accordingly, the activity, a, is an intensive parameter and is a direct measure of the difference between the chemical potential in the actual and in the reference state.

    In defining a chemical system it is important to distinguish between intensive properties based on chemical potential measurements and extensive properties based on counting the number of moles of a given substance. This can easily be illustrated in comparing data from potentiometric measurements of pH, pX or pM where X and M refer to anions and cations respectively, with those from titrimetric determinations of acidity, anions or cations. In the former case measurement is based on potential determinations which are essentially intensive parameters while

19

in the latter case measurement is based on stoichiometric calculation. Results of analysis of either type may not agree particularly if interferences are present which may cause the activity coefficient to deviate from unity, *e.g.*, salting in or salting out agents.

Similarly, in the case of the voltammetric membrane electrode systems, for measurement of dissolved oxygen, the measured parameter is essentially an intensive factor, since the diffusion current is solely dependent on the difference in the chemical potentials of molecular oxygen across the membrane. Accordingly, values derived from measurements with the galvanic cell oxygen analyzer do not have to equal results obtained by titration methods for dissolved oxygen, such as the Winkler test. In the former case, the activity of molecular oxygen is the parameter measured, while in the latter case, the total number of oxygen molecules present in the test volume is measured. For the majority of natural and wastewaters it is unlikely that the two kinds of measurements will give exactly the same results, although in many applications the differences may be negligible for all practical reasons. Examples of intensive and extensive measurements are shown in Table 6.

Table 6

Examples of Intensive and Extensive Measurement

Parameter	Measurement Intensive	Measurement Extensive
Proton condition	pH	Acidity or alkalinity
Electron condition	pE	Redox titrations
Hardness metals	pCa^{2+} and pMg^{2+}	EDTA titration method
Nitrates	pNO_3^-	Brucine titration method
Dissolved oxygen	a_{0_2} by M.E.	Winkler titration method

where $pH = -\log a_{(H^+)}$ $pNO_3 = -\log a_{(NO_3^-)}$

$pE = -\log a_{(e^-)}$ a_{0_2} by M.E. = oxygen activity by membrane electrodes

$pCa = -\log a_{(Ca^{2+})}$

Primary and Secondary System Characteristics

The most critical part of measurement systems is
commonly the sensor, and the reliability of measurement
is usually dependent on the reliability of this sensor
system. This is true whether the sensor is an electrode,
a thermistor or a photoelectric cell. A clear understanding of the operation characteristics of the sensor
and its dynamic response is essential. This is based on
proper calibration, servicing, maintenance and alertness
for small clues that may indicate malfunctions.

Primary sensor characteristics are defined in terms of
(a) sensitivity, (b) response time, (c) selectivity,
(d) long term stability, (e) accuracy, and (f) precision.
Secondary sensor characteristics are those which define
the environmental effects, for example (a) temperature,
(b) flow, (c) ionic strength, (d) pH, and (e) sunlight.

Primary sensor characteristics

Sensitivity is usually defined as the smallest change
in the measured variable that causes a detectable change
in the indication of the instrument. Sensitivity is
directly proportional to the slope of the curve relating
the signal magnitude to the amount of detectable material
present. This will reflect directly on the ability to
ascertain a difference between the signal and background
noise at the detection limit, *i.e.*, given adequate precision, the greater the sensitivity, the better the
detectability.

The limit of detection of an analytical method is the
lowest concentration whose signal can be distinguished
from the blank signal. It specifies the lower limit of
detection of the sensor. This value depends on the sensitivity of the method, as well as the signal-to-noise
ratio required to discern the response due to a sample.
Advances in electronics have brought about the design of
instruments with greater inherent stability and, therefore, lower limits of detection. Use of an on-line
digital computer in fast-sweep derivative polarography
has permitted the resolution of closely spaced peaks and
extended the analytical sensitivity of the technique by
more than an order of magnitude. For illustrative purposes
the detection limits of some of the more common methods
of analysis are given in Tables 7 and 8.

The speed of the sensor response to changes in the test
solution is referred to as the "response time." It is an
indication of the time needed for the sensor signal to follow 90%, 95% or 99% of instantaneous full scale change in the

Table 7

Detection Limits of Electroanalytical
Methods of Analysis

Method	Detection Limit
AC polarography, chronopotentio-metry, potentiometry	$10^{-4} - 10^{-5}$M
Classical polarography, coulometry at controlled potential, precision null-point potentiometry	$10^{-5} - 10^{-6}$M
Derivative polarography, square wave polarography, linear sweep voltammetry	$10^{-6} - 10^{-7}$M
Pulse polarography, amperometry with rotating electrodes	$10^{-7} - 10^{-8}$M
Anodic stripping with hanging mercury drop electrodes	$10^{-8} - 10^{-9}$M
Anodic stripping with thin film electrodes or solid electrodes	$10^{-9} - 10^{-10}$M

measured variable. The response time should be specified for each sensor, and this time must be indicated as either dynamic or static sensor response.

Selectivity of the sensor refers to the effect of interferences resulting from detectable ions or molecules other than the species of interest. Not all sensor systems can achieve absolute or 100% selectivity; hence, it is important to specify the selectivity limitations in a given test solution. If the type and amount of the interfering species are known, then it is possible to incorporate the term "selectivity coefficient" in the sensor's sensitivity expression. Also, in certain cases it is possible to incorporate interference effects in the sensor calibration curve. This can be done by means of the standard addition technique where known amounts of the measured ions are added to the test solution and the proportional signal values are recorded.

Table 8

Detection Limits of Spectrometric
Methods of Analysis

Method	Principle	Detection Limit
Molecular absorption spectrophotometry	Absorption of radiation by dissolved molecules	$10^{-5} - 10^{-6}M$
Molecular fluorescence spectrophotometry	Re-emission of radiation absorbed by dissolved molecules	$10^{-7} - 10^{-8}M$
Atomic absorption spectrophotometry	Absorption of radiation by free atoms	$10^{-6} - 10^{-7}M$
Atomic fluorescence spectrophotometry	Re-emission of radiation adsorped by free atoms	$10^{-7} - 10^{-8}M$
Optical and X-ray spectroscopy	Spectral emission analysis by flame, arc or X-ray excitation	$10^{-5} - 10^{-6}M$
Neutron activation analysis	Nuclear activation by thermal neutrons	$10^{-9} - 10^{-10}M$

Long term stability usually refers to the change in the sensor's performance characteristics with time. This is used to decide on the frequency of checking the calibration or servicing the sensor. Long-term stability is a property of the particular system and is dependent on the presence of interferences and the physicochemical characteristics of the test solution.

Deviations of results by a given sensor from the "true" value define the accuracy of the system. If the source of error is found, and it is possible to correct for, this is called "determinate error." If the deviation from the true value is compounded indiscriminately by many small

23

errors, it is simply a "random error." Random errors are subject to statistical treatment of the data.

Precision is defined in terms of the reproducibility of the sensor measurement. The more scatter in successive readings, the less precise are the measurements. Usually, precision is closely identified with random errors and statistical theories.

An important distinction between precision and accuracy is that accurate measurements are always precise, but the converse is not necessarily true. The precision of a series of measurements may be good, but every result may be higher than the true value because of an unsuspected interference. Only when determinate errors are minimized will precision also imply accuracy.

Secondary sensor characteristics

Secondary sensor characteristics refer to the effect of environmental variables. This can be a result of changes in the sensor's primary characteristics or changes in the physicochemical characteristics of the test solution. For example, temperature effects on conductance measurement are quite complex since the temperature coefficient is dependent on both ionic strength and temperature. The conductivity of sea water was found to increase by 3% per degree increase in temperature at 0°C, 2% increase at 25°C and about 5% increase at 30°C. It is therefore advisable to measure relative conductance rather than absolute values.[4] This is done by measuring the ratio of the conductance of the test solution to that of a reference solution at the same temperature. Thermistors or resistances can be used instead of the reference solution.

It is always advisable to establish the primary and secondary sensor characteristics for each sensor independently before using it for field applications. Variations may occur between sensors of different types and also between sensors of the same type and from the same manufacturer.

REFERENCES

1. Federal Water Quality Administration "Annual Record of Water Quality" Surveillance Program Papers 1962-1971.
2. U. S. Geological Survey "Annual Record of Chemical Quality" Water Supply Papers, six volumes (1954-1959).
3. Beeton, A. M. Limnol. Oceangr. 10, 240 (1965).

4. Federal Water Quality Administration, "Lake Erie
   Report k-A Plan for Water Pollution Control,"
   Department of the Interior (Washington, D.C.: 1968).
5. Kopp, J. F., and R. C. Kroner. "Trace Metals in
   Waters of the United States," Federal Water Pollution
   Control Administration (Cincinnati, Ohio: 1965).
6. Lohr, E. W., and S. K. Love. "The Industrial Utility
   of Public Water Supplies in the United States, 1952,"
   Parts I and II, Geological Survey Water-Supply Papers
   1922 and 1300 (Washington, D.C.: 1954).
7. Love, S.K. "Quality of Surface Waters for Irrigation,
   Western States, 1963," Geological Survey Water-Supply
   Paper 1952 (Washington, D.C.: 1967).

Chapter III

Design of Sampling Programs

Sampling Programs

The goals of any study must be established before a
valid program of sampling, analysis and interpretation can
begin. A satisfactory sampling program for a water pollution study can then be designed with the assistance of
scientists from all disciplines involved. Preliminary information including the number of outfalls, their locations, and the types and quantities of wastes being
discharged will greatly simplify the process of designing
a sampling program. The variability of the constituent in
the environment to be studied should be determined so that
the frequency of sampling can be determined for the desired precision.

No general procedures for the establishment of a
satisfactory sampling program applicable to all situations
can be stated. The composition of wastewater or natural
water is dependent upon many factors as indicated in the
previous chapter. To satisfy the requirements of many
studies, only a few samples may be required. This is
often the case in surveys of industrial effluents. For
other studies frequent sampling at many locations may be
required.

In any study, certain basic criteria are required in
the sampling program. To describe the environment by the
use of such statistics as a mean and variance of a finite
number of samples, a valid sampling program must be
adopted. Samples should be randomly collected to permit
statistical evaluation. In practice, random sampling is
infrequently used; instead it is common procedure to

sample at regular intervals. This always leads to the possibility that the results will be biased because of a cyclic fluctuation in concentration which is in phase with the sampling program. Such bias will not occur if the samples are collected with a frequency greater than the cyclic variation of the environment. Dissolved oxygen, for example, undergoes a diurnal variation. Photosynthesis of phytoplankton during daytime may result in supersaturation of oxygen in the surface water of lakes, while at night respiration may result in undersaturation of oxygen. It is possible to describe this diurnal variation by collecting oxygen samples at regular intervals which are short in comparison to its cyclic variation. If, however, samples are collected daily, the data may be incorrectly interpreted.

The number of samples required to describe a natural water depends on the variability of the constituent to be analyzed. The sampling program should be flexible enough to permit changes if this seems desirable. The total number of samples collected from all locations is governed by the facilities of the analytical laboratory. Too frequently the samples are divided among too many stations. The collection of a larger number of samples at fewer stations permits statistical analysis of the data with much more reliable results.

Factors which should be considered in the selection of sampling sites have been discussed by Velz.[7] During periods of relatively steady flow the data will be much easier to interpret than during periods of highly varying flow. In general, sampling of streams should be conducted downstream from outfalls and tributaries at a point where dispersion throughout the stream is complete. Sampling too close to these discharges may result in erratic results. If sampling cannot be conducted at a point where complete mixing has taken place, samples should be taken from the stream above its mouth. Physical characteristics of some rivers may require the collection of samples at many points across the channel and at different depths. Analyses of streams which receive industrial waste discharges will reflect the fluctuations in these discharges. Therefore, in planning a sampling program consideration should be given to the operating schedule and waste discharge patterns of industries influencing the river. Likewise, river levels may be greatly affected by diurnal variations in the storage and discharge of water, especially by hydroelectric plants.

Sample Collection

Most water analyses are performed on grab samples collected in plastic buckets or such samplers as Kemmerer, Van Dorn and Nansen bottles. Samplers which are programmed to collect a volume of sample at regular intervals are available. These devices generally produce a composite sample, therefore losing much of the data which would result from the analyses of individual samples. Continuous analyses depend on the sensors being immersed in the stream or on water being pumped by the sensors. If water is pumped, flow rate and tubing size should be chosen to minimize the time the sample is in the tubing.

The sampling procedure must be capable of providing a valid sample. Care should be exercised in the collection of samples for the analysis of dissolved gases. Usually a satisfactory sample can be obtained by slowly drawing water from the sampler into the sample bottle through a tube extending to the bottom of the bottle and allowing water to overflow until two to three times the volume of the bottle has been allowed to overflow. If samples are to be analyzed for trace metals, nonmetallic samplers should be used to prevent sample contamination.

An important part of any sampling program is the provision of complete information on the source and conditions under which the samples were collected. The United States Geological Survey[5] recommends that the following data be collected for each sample:

Surface Waters	Ground Waters
Name of water body	Geographical and legal
Location of station or	locations
site	Depth of well
Point of collection	Diameter of well
Date of collection	Length of casing and position
Time of collection	of screens
Gauge height or water	Method of collection
discharge	Point of collection
Temperature of the water	Water bearing formation(s)
Name of collector	Water level
Weather and other natural	Yield of well in normal
or other man-made factors	operations
that may assist in	Water temperature
interpreting the chemical	Principal use of water
quality	Name of collector
	Data of collection
	Appearance at time of
	collection

(cont)

Surface Waters	Ground Waters
	Weather or other natural or man-made factors that may assist in interpreting chemical quality.

Sample Preservation

Analyses are of little value if the sample has undergone changes between sampling and analysis. Some analyses require lengthy procedures or specialized equipment which often cannot be adapted readily to field use. The analyst must therefore rely on samples which have been stored for some period of time. Changes in samples may be a result of physical, chemical, or biological factors, but, since all are time dependent, the shorter the storage time, the less the effect on the sample.

Some sample constituents change so rapidly that they must be measured at the sampling site. Changes in temperature and pressure will cause the concentrations of dissolved gases (*e.g.*, O_2, CH_4, H_2S, and CO_2) to change. It is possible to "fix" some components through appropriate chemical treatment. Sulphide may be stabilized for many hours by formation of a mixed zinc sulfide-zinc hydroxide precipitate.[1] Shifts in the carbonate or sulfide equilibria through the release of gas or precipitation will cause the pH to shift and it is therefore recommended that pH measurements be made at the sampling site.

Samples for heavy metal ion analysis should be filtered at the sampling site and acidified to about pH 3.5 with glacial acetic acid.[5] Acidification minimizes precipitation and adsorption onto the walls of the container. Acetic acid stimulates the growth of molds, sometimes making it necessary to add a small amount of formaldehyde to the sample as a preservative.

Strickland and Parsons[6] recommend storing samples at -20°C for nitrogen and phosphorus if the analyses must be delayed for more than a few hours. Samples for nitrogen analysis should not be acidified since acid will act as a catalyst in the Van Slyke reaction

$$O=N-O^- + \underset{\underset{NH_2}{|}}{RCHCOOH} + H^+ \rightarrow \underset{\underset{OH}{|}}{RCHCOOH} + N_2 + H_2O$$

and low values will be obtained. Mercuric chloride has been recommended as a preservative for inorganic nitrogen analyses.[2,4]

Because of biological utilization of orthophosphate and adsorption of phosphate onto container walls, orthophosphate may be lost from solution during storage. If samples are acidified, hydrolysis of polyphosphates and organic phosphates may result. Heron[3] found that in lake water with low concentrations of phosphate (less than 10 ppb PO_4-P) reduction of phosphate was due to bacterial action rather than adsorption. By treating a clean polyethylene bottle with a 5% solution of iodine in 8% potassium iodide for one week, samples could be stored for as long as two weeks without appreciable change.

Corollary Field Data

Concurrent with the collection of samples for subsequent chemical analysis or with continuous monitoring of chemical constituents, most programs will require other measurements to be made or samples to be collected. Many of these measurements may be made by the chemist or may be of value to him in the evaluation of chemical data. The measurements which are physical in nature are discussed in Chapter 9. In most laboratories these will be the responsibility of the chemist. Biological measurements of bacteria, plankton, and benthos will be useful in the assessment of pollution, and close coordination of biological and chemical programs should exist for nutrient studies.

Stream flow should be measured at the time samples are collected. This information is essential to any well-planned sampling program since it will determine not only the degree of dilution of wastes but also the quantity of material contributed by run off.

REFERENCES

1. American Public Health Association, "Standard Methods for the Examination of Water and Wastewater," 12th ed. (New York: 1965) p 293.
2. Brezonik, P. L., and G. F. Lee. Air Water Pollut. 10, 549 (1966).
3. Heron, J. Limn. and Oceanog. 7, 316 (1962).
4. Jenkins, D. "Trace Inorganics in Water," Advances in Chemistry, No. 73, R. F. Gould, Ed., Am. Chem. Soc. (Washington, D.C.: 1968) p 265.

5. Rainwater, F. H., and L. L. Thatcher. "Methods for Collection and Analysis of Water Samples," U. S. Geological Survey Water-Supply Paper 1454 (Washington, D.C.: 1960) p 20.
6. Strickland, J. D. H., and T. R. Parsons. "A Practical Handbook of Seawater Analysis," Fisheries Research Board of Canada Bulletin 167 (Ottawa, Canada: 1968) p 82.
7. Velz, C. H. Sewage Ind. Wastes 22, 666 (1950).

Chapter IV

Concentration and
Separation Techniques

Introduction

When one analyzes aqueous systems, the need frequently
arises to remove interferences or to improve sensitivity.
The former may be achieved by a variety of separation
techniques, while the latter may require the use of concentration methods. Indeed, the essence of several
analytical techniques is a concentration or separation
procedure. Thus, the widely used carbon adsoption method
for monitoring water quality essentially involves a concentration procedure, while the basis of gas chromatographic analysis is a separation process. Similarly, the
membrane filtration procedure in the enumeration of
bacteria in water involves and may be regarded as either
separating the bacteria from the water or concentrating
them within it.

Aside from the need for separation in order to remove
interferences, it is frequently useful to distinguish
between soluble and particulate species in water. One
such means is using the separation technique of filtration
by a membrane with a nominal pore diameter of 0.45
microns. This procedure does not perform a perfect separation of particles larger and smaller than the nominal
pore size, due to both the entrapment of smaller colloidal
species as well as passage of some larger species because
of heterogeneity in pore size. Nevertheless, it is an
operationally useful technique in that commercial,
relatively reproducible, and uniform membranes are available, so that results among various investigators should
be repeatable.

This chapter will discuss five basic concentration and
separation techniques in some detail: carbon adsorption,

freeze concentration, liquid-liquid extraction, ion exchange, and chromatography (particularly gas-liquid chromatography). Other techniques will be mentioned briefly. For a more comprehensive discussion of concentration and separation techniques the reader is referred to Berg.[9] Andelman and Caruso[3] have recently reviewed the principles of several such methods and their applications to water analysis.

Carbon Adsorption

The sorptive properties of activated carbon and its particular effectiveness for removing certain organic materials which impart tastes and odors to water supplies have been widely recognized. In the last five years, a specific sampling procedure using activated carbon has been recommended by the U. S. Public Health Service and currently enjoys a 'tentative method' status.[2] The carbon filter provides a direct means of concentrating trace organic compounds in water through passage of large volumes (as much as several hundred thousands of gallons) of water through a column of 30 mesh activated carbon (Nuchar C-190, West Virginia Pulp and Paper Co., or equivalent).

After an adsorption run during which the total volume of sample flow (0.25 gpm) has been measured, the carbon is removed from the column and then dried and extracted with chloroform, a solvent which recovers materials most likely to be responsible for tastes and odors. Included, however, are oils, phenols, various synthetics and other materials of slight water solubility. Chloroform extraction is followed with an alcohol extraction, which often removes even more organic material than was recovered with chloroform. Alcohol extracts are apparently of a different nature and usually do not have the intense odors exhibited by the chloroform solubles. Both solvent extracts are concentrated to small volumes on steam baths and then air dried overnight to reduce loss of volatile material. Final weights of the extracted materials are determined and the data recorded as:

$$\text{carbon - chloroform - extract (CCE)} \ \mu g/l = \frac{g \ CCE \times 10^6}{gal \ sample \times 3.785}$$

$$\text{carbon - alcohol - extract (CAE)} \ \mu g/l = \frac{g \ CAE \times 10^6}{gal \ sample \times 3.785}$$

According to the U. S. Public Health Service[49] a concentration of 200 $\mu g/l$ of CCE represents excessive organic contamination.

The extracts are complex mixtures and little can be determined regarding their chemical composition without further separation. Solubility fractionation as outlined in Figure 1 may be employed to further characterize chloroform or alcohol solubles. The Ether Insolubles are tarry or polymerized substances which do not appear

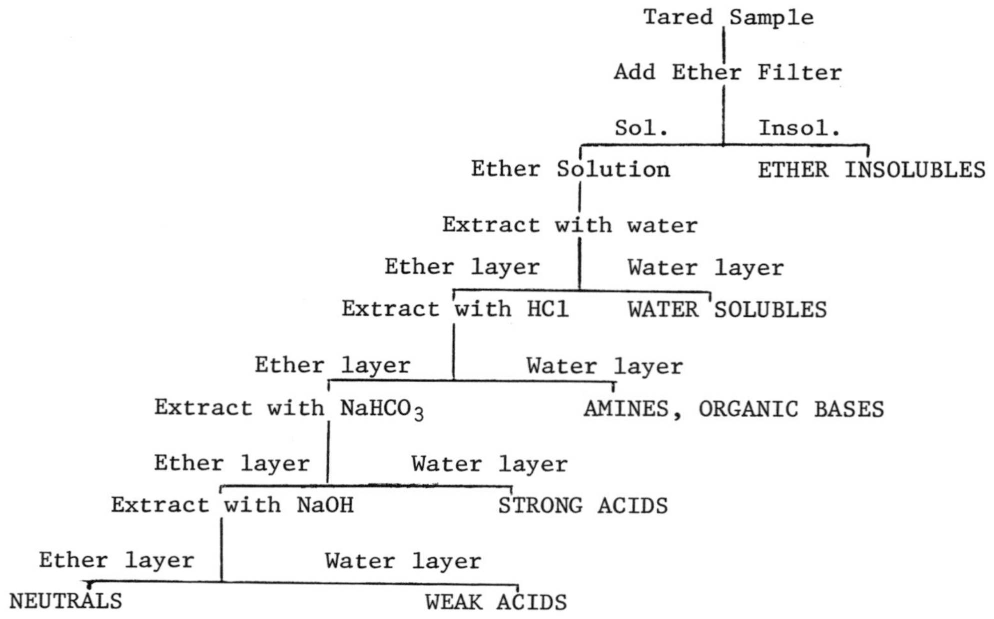

Figure 1. Solubility Separation of Extracted Organic Material

important as taste or odor producers. They are removed primarily to prevent interference in later analytical steps. The Water Solubles apparently contain humic-type materials of natural origin or derived in part from sewage. The Weak Acid fraction will contain phenols and the weaker carboxylic acids and sulfonic acids. The Strong Acids are often quite odorous (fruity or rancid), their typical components being acetic, butyric and caproic acids. These substances result from biological action on sewage, industrial material or organic debris. Materials which are Neutral to the acid-base solubility fractionation can be separated on columns of silica gel by selective elution with iso-octane (aliphatic fraction), benzene (aromatic fraction) and chloroform-methanol (oxygenated fraction) and recorded as weight percentages of a given fraction. One might expect to find polynuclear hydrocarbons of the benzo-(a)-pyrene type or substances like DDT in the aromatic subfraction of the Neutrals solubility category.

Recoveries of organic material from the CCE and CAE procedures can be increased in some cases by removing sample turbidity with diatomite filters prior to contact with the carbon bed.[46] Increased extraction yields with chloroform and alcohol are obtained if extraction of the carbon is performed at pH 3.0. Extensive research has been performed on the mechanisms of the carbon adsorption of many compounds of special interest in water supply or waste treatment.[50]

Freeze Concentration

The use of freezing techniques to concentrate water which contains organic material is attractive due to the structural protection offered by reduced temperatures. As solute-laden aqueous samples are frozen, the ice crystals which form are extremely pure because solute is rejected to the liquid phase. If total solidification is prevented, the residual liquid will be greatly enriched in all solutes.

Freezing may be accomplished by rotating a round bottom flask at a controlled rate in a temperature regulated bath.[5] Baths of crushed ice and salt or ethylene glycol-dry ice are commonly used. Sample volumes of 200 ml to 20 liters may be handled in batch fashion, although it is generally more convenient to freeze multiple samples of smaller volume.

Although many factors theoretically affect the efficiency of solute concentration during freezing (*i.e.*, degree of mixing, freezing rate, nature and concentration of solutes), Shapiro[43] has contended that 99% recovery can be obtained from dilute organic solutions at volume ratios of 20:1; since losses are nonspecific, overall recoveries could be determined by conductivity measurements. Other workers,[6,24] notably Baker, have shown that specific organic recoveries may not be in agreement with recoveries measured by conductivity and that the process may be limited by ionic concentration in the residual liquid. Most recently, Baker[5] has studied the recoveries of several known chemical structures by freezing and observed that the mixing rate is not a factor in the absence of dissolved organic salts. He also notes that neither molecular weight and size nor the nature and location of substituent groups affect recoveries in the absence of inorganic solutes. In the presence of inorganic salts, organic recoveries are markedly reduced although higher mixing rates are helpful. It is interesting and important to note that this loss of efficiency is apparently not structurally selective. Thus, as Baker has recommended, the concentration of all organic components may be adjusted according to the observed concentration efficiency of an added standard.

Some workers[14] have concentrated trace organics by a combination of vacuum distillation at temperatures not exceeding 50°C and dialysis and lyophilization (freeze drying). The lyophilizer apparatus consists of two stainless steel cylinder halves. The top half (2" x 25") has 48 attachment ports welded to the cylinder. The bottom half (3" x 10") is a water trap designed so that the air travels down through the center to the bottom of the cylinder, then up along the outside wall to an exit port which is connected to a vacuum pump. A Dewar flask containing dry ice in acetone (-70°C) is placed around the outside of the lower cylinder to provide a coolant for the water trap. The sample is placed in a thick-walled vacuum filter flask which is then capped with a rubber stopper and joined to one of the attachment ports with vacuum tubing. All joints are sealed with a suitable high-vacuum grease. The system is then evacuated.

The overall process is relatively slow, but it leaves a dry solid residue of organic matter (high molecular weight) which is relatively ash free and can be easily re-solubilized in water.

Liquid-Liquid Extraction

Liquid-liquid extraction (to be referred to here as

solvent extraction) has been widely used in water analysis for separating and concentrating a great variety of materials, including organic and inorganic ions, and neutral species as well. It can be performed relatively simply and inexpensively but is also adaptable to more expensive and elaborate multistage and countercurrent techniques which can simultaneously increase the concentration factor and separation efficiency.

Solvent extraction avoids exposure of the extracted species to heat or solid surfaces, which may result in reactions or structure changes for labile organic materials. Unlike precipitation techniques for inorganic ions, which can result in coprecipitation of unwanted species, it can be highly selective and can be used for very small quantities of material.

Principles

In order to perform a solvent extraction of an aqueous species, it is necessary for the extracting solvent to be essentially immiscible in the water, and vice versa. In practice there is always some slight miscibility on both parts. The extracting solvent is shaken with the aqueous solution for a time sufficient for the extractable solute to equilibrate between the two phases; the phases are then permitted to separate and the solution of the extracting solvent containing the extracted solute is removed from contact with the water and utilized as required, either for further separation and concentration or directly in analysis. The smaller the volume of the extracting solvent utilized in this process, the greater will be the concentration factor, but the smaller will be the recovery.

The distribution equilibrium between a solute that is nonionic and is in the same molecular form in the two phases (*i.e.*, water and extracting organic solvent) is essentially equal to the ratio of its solubilities in the two phases. This holds either when the solutions are ideal or are dilute enough to behave in an ideal fashion. For such a nonionic solute A distributed between water, designated by "W," and an organic solvent, designated by "O," the equilibrium ratios of concentrations may be expressed as

$$P = [A]_O/[A]_W \tag{3}$$

which is the Nernst partition law; P is defined as the partition coefficient.[26] The partition coefficient for many species, particularly those which are charged, are

39

dissociate, or are in different forms of aggregation in the two liquid phases, can vary considerably with concentration, changes in pH, and the addition of salt or complexing agents.

The fraction of material originally present in the water phase and extracted by the organic solvent may be derived from Equation 3 to give

$$F_E = 1/[1 + V_W/(V_O P)] \qquad (4)$$

with F_E as the fraction extracted and V_W and V_O the volumes of the water and extracting organic solvent, respectively. Thus, the greater the ratio V_O/V_W and the greater the value of P, the greater will be the fraction of material extracted from the water into the organic solvent phase, approaching unity in the limit. In order to increase the recoverability of a species in solvent extraction, it may be necessary to perform a series of extractions on the water, using a fresh solvent each time. With constant values of P and using the same volume of organic solvent in each of n successive extractions, the total fraction extracted is then

$$F_E \text{ (multiple)} = 1 - (1 - F_E)^n \qquad (5)$$

Thus, for example, if $F_E = 0.9$ for one extraction, for two successive extractions $F_E = 0.99$. Both multistage[15] and continuous countercurrent extractors[25] have been developed which make use of this increased recovery with exposure to fresh solvent.

Another important consideration is the concentration factor, the ratio between the concentration of extracted material in the organic phase at equilibrium, compared to that *initially* present in the water solution. Using Equation 3 this may be shown to be

$$\text{Concentration factor} = 1/(1/P + V_O/V_W) \qquad (6)$$

40

Thus, the concentration factor will increase with the magnitude of the partition coefficient but decrease with increasing V_O/V_W.

In extracting inorganics, such as metal ions, from water, a common technique is to use an organic complex or chelating agent which greatly increases the solubility of the metal in the organic solvent, thereby increasing its partition coefficient.[26] In such an extraction, both the uncharged metal-chelate molecule and the organic chelating agent itself have separate partition equilibria. Because these chelates are generally weak acids or bases, pH will have a great effect on their ability to bind the ion, as well as on their own partition coefficient. The combination of these effects is such as to cause pH to have a high degree of control in the extraction of metals by organic solvents using chelating agents. This can be very useful in selectively extracting certain metals in a mixture.

Applications

In one recent application solvent extraction was used to concentrate chlorinated hydrocarbon pesticides from water using a semiautomatic device.[22] Following extraction of 850 ml of water by 50 ml of various solvents, the latter solutions were then evaporated to 0.5 ml prior to gas chromatographic analysis. Compared to a manual extraction process this semiautomatic technique was about 50% more efficient, with recoveries varying from 69 to 95% in the combined steps of extraction and evaporation.

In an investigation on taste- and odor-producing organics in river and municipal water, an 18-stage continuous countercurrent extractor was used to concentrate prior to analysis.[19] Several solvents were tested, and it was concluded that methyl isobutyl ketone was the best extractant for phenol and was generally good for other organics. Using an aqueous feed of about 0.1 mg/l of phenol, recovery was 94%. Extractions of mixtures containing o-cresol, guaiacol and phenol led to recoveries of 88 to 99%, depending on the solution-solvent ratio and agitator speed. In using this technique to extract river water under the same conditions used for municipal water, the sample-solvent mixture tended to emulsify, and it was necessary to reduce the agitator speed. In one such extraction of river water containing 0.07 mg/l of phenol, the recovery efficiency of the latter was 52%.

Many analyses of metals involve their chelation,

followed by solvent extraction. One recent study was made
of a comparison of the extraction from sea water of
iron(II) complexes with orthophenanthroline, bathophenanthroline, and 2,4,6-tripyridyl-symtriazine, the
extracting solvent being propylene carbonate.[48] It was
found that, using these chelates, iron(II) was readily
extracted from sea water at low concentrations and could
then be analyzed in the propylene carbonate spectrophotometrically. For iron(II) in the concentration range of 5
to 27 µg/l, the average error was 5%. Depending on the
chelate used, the pH range of extraction was from 2 to 9.

Solvent extraction of metals is frequently used before
atomic absorption analysis, not only to concentrate but
also to utilize the increased sensitivity of the method
with organic solvents. A common solvent for this purpose
is methyl isobutyl ketone (MIBK). One study using this
solvent involved the extraction of cobalt, nickel and lead
in fresh water, using ammonium pyrrolidine dithiocarbamate
as the chelating agent.[16] In these studies the water-MIBK
ratio was 20/1 and the pH in the water adjusted to 2.8
prior to extraction. This pH value was critical for lead,
but could be as high as 5 for cobalt and nickel. It was
concluded that this extraction technique in combination
with atomic absorption spectrophotometry offered a rapid,
simple, accurate, and sensitive method for analyzing these
metals in fresh waters. A similar method was studied for
iron, copper, zinc, lead and cadmium, using diethyldithiocarbamate as the chelating agent.[35] As a result of the
concentration of metals in the extraction process and the
increased sensitivity of the atomic absorption analysis in
the MIBK, the detection limit was lowered by a factor of
15 to 30 for each metal, compared to its direct analysis
in water.

These few examples demonstrate part of the range of
applicability of solvent extraction as a tool to facilitate water analysis. It is a widely used technique, both
for concentrating and separating a great variety of micro
and macro species in water.

Ion Exchange[1]

Ion exchange as a concentration and separation technique has been in use for some 50 years. One early method by Bahrdt in 1927 used a sodium zeolite column to remove calcium and magnesium from natural water because of their interference in a sulfate analysis.[41] The use of synthetic zeolites in ion exchange is limited, however, because of their narrow useful range of pH and the difficulty in achieving quantitative elution. In the 1930's, organic cation and anion exchange resins were synthesized and analytical applications developed, principally by Samuelson.[40] The theory of ion exchange equilibria, kinetics and chromatography has been treated by Helfferich,[18] and Samuelson[41] and Inczedy[20] have comprehensively considered their analytical applications.

Ion exchange in water analysis may be used to determine the total equivalents of salt present, to concentrate ions, and to separate them from nonelectrolytes as well as from other ions of similar or opposite charge. It is applicable to organic and inorganic ions, and is particularly useful in removing interferences in order to facilitate subsequent analysis. Ion exchange is a fast and simple technique which usually yields high accuracy and recovery and requires relatively little judgment, thus making it readily adaptable for routine analyses.

Ion Exchange Equilibria and Kinetics

When an ion exchanger is equilibrated with an ambient solution containing two exchangeable ions, univalent cations for example, the exchange process is stoichiometric and the equilibrium reaction is

$$A^+ + B_R^+ \rightleftharpoons A_R^+ + B^+ \tag{7}$$

with the subscript R referring to species in the resin phase. (The term "resin" will be used for convenience to designate the ion exchange material, since synthetic resins are the principal ion exchange material in current use.) The absence of this subscript refers to species in the ambient solution phase. For this reaction an equilibrium expression may be written

---

[1]With the permission of the publisher, portions of this discussion are taken from the article by J. B. Andelman and S. C. Caruso, "Concentration and Separation Techniques" in *Handbook of Water and Water Pollution*, L. Ciaccio, Ed., (New York:Marcel Dekker, Inc., in press).

$$K_B^A = \frac{(A^+)_R \times (B^+)}{(A^+) \times (B^+)_R} \tag{8}$$

with K_B^A being the equilibrium constant, which is generally referred to as the selectivity coefficient. The terms in parentheses refer to concentrations. K_B^A is a simple measure of the ability of the resin to select A^+ over B^+. For example, if in the ambient solution at equilibrium $(A^+) = (B^+)$, then $K_B^A = (A^+)_R/(B^+)_R$. The larger the selectivity coefficient, the greater is the efficiency of separating the ions of like charge by ion exchange.

Selectivity scales for various ions and resin types have been constructed. They indicate the relative affinities of a given resin for various ions. For example, with a typical strong acid resin an affinity series for univalent cations is[41]

Ag > Cs > Rb > K > NH$_4$ > Na > H > Li.

A similar scale for divalent ions with the same type of resin is

Ba > Pb > Sr > Ca > Ni > Cd > Cu > Au > Mg.

The selectivity coefficient for a given pair of exchanging ions is affected by several factors. One of the most important is the basic chemical structure of the resin. When, for example, the cation exchanging sites are of the strong acid type, such as sulfonate, the resin affinity for hydrogen ion is generally low. In contrast, a weak acid cation exchange resin, that with carboxyl sites, has a high affinity for hydrogen ion. Other factors that affect the selectivity coefficient are composition of the exchange ions in the resin phase, tightness of the resin structure, temperature, and ambient solution concentration of exchange ions and other solutes.

Most of the ion exchange sites in a typical resin are located within the matrix or pore structure. Thus, for a typical ion exchange process to occur, such as represented by Equation 7, the ions must pass through the resin matrix and across a liquid film boundary layer at the resin solution interface. The rate determining step in the exchange process could then be[11]

(A) diffusion in the boundary layer (film diffusion),
(B) diffusion in the resin phase (particle diffusion),
or (C) chemical exchange at the exchange sites.

In most cases, the chemical exchange is rapid enough that
it is not rate limiting. In one study of alkali metal ion
exchange it was found that at ambient solution concentrations below 0.003M the exchange kinetics could be
considered to be film diffusion limited; above 0.1M the
process was particle diffusion limited.

Column Operation

The rate of the exchange process and the selectivity
coefficient are the principal factors that affect column
operations. The latter are widely used because they lend
themselves readily to continuous operation, they can be
used in chromatographic separations, and the exchange
reaction approaches completion because it is continuously
displaced.

If the primary purpose of ion exchange is to concentrate either a single ion or a mixture of similarly
charged ions, the solution may be passed through the
column. The resin behind the advancing front is left
with, for example, the cation to be concentrated,
A^+, or a mixture of such ions A_1^+, A_2^+, A_3^+, etc. At this
point the column may be washed with distilled water, the
front remaining fixed in place. Next the column is
eluted with an electrolyte, perhaps containing C^+, and the
adsorbed A^+ ions move down the column ahead of the newly
advancing C^+-A^+ front. If the concentration of the
eluant C^+ solution is significantly larger than the
original A^+ solution, then the latter will appear more
concentrated in the effluent than originally. This procedure also serves to remove either anions or neutral
species originally present in the A^+ solution, because
when A^+ is initially adsorbed by the resin, they continue
through the column and appear in the effluent; any remaining quantities of anions or neutral species are
generally removed in the distilled water wash prior to
elution.

When ion exchange is being used to separate ions of like charge sign, it is useful to adsorb the mixture, but utilizing only a small portion of the top of the column. This is generally followed by washing the column to remove any excess co-ions and neutral species, and finally eluting by either the technique of selective displacement with various eluants or elution chromatography.

In elution chromatography, the mixture of similarly charged ions at the top of the column is eluted by one eluant which displaces all of them, but at different rates depending on their relative selectivities. As the eluant moves down the column, the bands of the various ions being eluted move with it, the peaks broadening as they move. The positions of the bands of such a mixture of three cations, A_1^+, A_2^+, and A_3^+, at two different times in the elution process are shown in Figure 2B. In this case A_1^+ and A_2^+ are not completely separated. Thus, although A_3^+ may be completely collected in one or more fractions not containing the other A^+ ions, A_1^+ and A_2^+ cannot. The similar movement of a single eluted substance is shown in Figure 2A.

Applications

There have been numerous reported applications of ion exchange in water analysis, only a few of which will be discussed briefly in order to indicate their scope. The total salt concentration in natural and boiler waters has been determined by adsorbing the latter onto cation exchange resins in the hydrogen form and titrating the displaced hydrogen ions. In one such technique a batch equilibrium method was used, with corrections being made for alkalinity which was determined separately, and the expected average deviation was 0.12 milliequivalents per liter of total cations. It was noted that for natural waters low in potassium and calcium, the total cation content is a good estimate of sodium.[33]

A scheme of analysis for industrial waters has been presented using strong acid and strong base resins in order to determine calcium, magnesium, copper, iron(III), chromium(III), chloride, sulfate, metaphosphate, orthophosphate, silicate and chromate.[30] The method was developed so as to remove chromate interferences in the colorimetric determinations of orthophosphate, silicate and metaphosphate, and in the chelometric analysis of calcium and

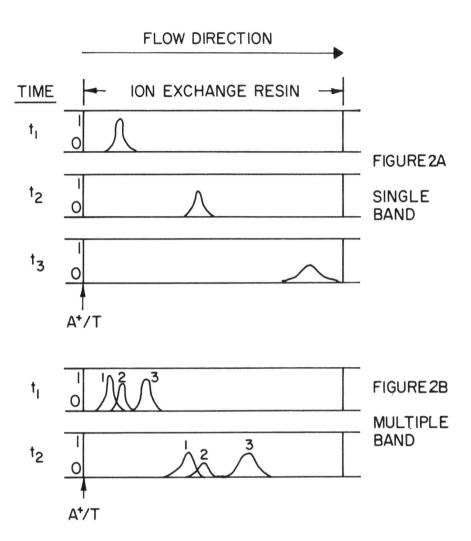

FLOW DIRECTION

TIME ⊢— ION EXCHANGE RESIN —⊣

FIGURE 2A

SINGLE BAND

A⁺/T

FIGURE 2B

MULTIPLE BAND

A⁺/T

Figure 2. Ion Exchange Elution Process. Reprinted from Reference 3 by Courtesy of Marcel Dekker, Inc.

magnesium. It also removed copper and iron(III) which interfered in the latter analysis. In the analysis of the anions, hydroxylamine hydrochloride is added to the test solution to reduce the chromate ion to chromium(III),

47

which is then removed in subsequent passage through the
cation exchange resin. The anions in the effluent are
then analyzed colorimetrically. Similarly chromium(III)
in the test solution is oxidized with peroxide to chromate
and separated from the cations by adsorption on the anion
exchange resin. Copper and iron(III) are then determined
directly in the effluent. For the calcium and magnesium
analyses, the copper and iron(III) are then complexed with
cyanide to form $Cu(CN)_6^{-2}$ and $Fe(CN)_6^{-3}$ which are removed in
passage over an anion exchange resin, the effluent then
being analyzed by EDTA titrations. The complete scheme of
analysis generally gives results which agree well with
conventional methods of analysis of cooling water samples.

Using a strong acid and a weak base resin column in
series, a field procedure was developed to enrich and
analyze natural waters.[34] Following uptake on the columns,
the latter were eluted by hydrocholoric acid and ammonia,
respectively. It was found that uptake and recovery were
complete for sodium, potassium, magnesium, calcium,
manganese(II), chloride, and sulfate. Variable amounts of
phosphate and iron(III) were adsorbed onto the resin,
the remainder passing through as nonexchangeable complexes
or complexes associated with humus.

A scheme for the analysis of the major cations in sea
water has been developed. It uses a resin column to
adsorb the cations and then successively elutes with a
sequence of different eluants.[17] After adsorbing 30 ml of
sea water on the resin, potassium and sodium were eluted
with 0.15M ammonium chloride. Then calcium was eluted
with a solution of 0.35M ammonium chloride, next magnesium
with 1M ammonium acetylacetonate at pH 9.6, and finally
strontium with 2M nitric acid. A variety of analytical
methods were used for the different fractions.

Several applications of ion exchange, used primarily
for the purpose of concentrating ions, have been reported.
A chelating resin has been used with industrial waste
waters in order to concentrate multivalent cations by a
factor of 10 to 20 before analysis by atomic absorption
spectrophotometry.[10] It was noted that the metals were
held more strongly by the chelating resin than by a
typical sulfonic acid resin. In this technique the water
samples were buffered to a pH of 5.5, sorbed onto the
resin, and then eluted with 8M nitric acid. The sensitivity of the method was 5 ppb for copper, cadmium and
zinc, and 50 ppb for lead(II), nickel(II) and iron(III).

One of the most prevalent uses of ion exchange in
water analysis is the removal of interfering species. One
such method for fluoride involves the use of an anion

exchange resin.[23] The test solution is passed through the resin column in the acetate state; the fluoride is retained, and potentially interfering cations appear in the effluent. The fluoride is then eluted from the column with 0.005M beryllium(II) in 0.1M acetic acid, the eluted species presumably being BeF_4^{-2}, and the SPADNS method used to analyze the fluoride. If aluminum is present in the test solution it must be chelated prior to the ion exchange step. Using synthetic fluoride solutions of 0.5 and 1.0 mg/l, the accuracy and precision of the method were each approximately 10 per cent, even with the following added possible interferences: 200 mg/l orthophosphate, 400 mg/l calcium carbonate, 1000 mg/l chloride, 2 mg/l hexametaphosphate, 1000 mg/l sulfate, 0.5 mg/l chlorine, and 5 mg/l aluminum. As with the previous cation exchange method, this technique eliminated the need for distillation.

Ion exchange is also a useful tool in the analysis of organic constituents in natural waters. An example of such an application is the analysis of paraquat, which is the common name for the cation 1,1-dimethyl-4,4-bipyridilium, used as a herbicide for weed control.[12] Plant constituents and other organics in water may interfere with the spectrophotometric analysis and are removed by this ion exchange technique. After passing the aqueous test solution through the cation exchange resin column, the latter is washed first with 2N HCl, then 2.5% NH_4Cl. The paraquat is then eluted with saturated NH_4Cl and analyzed spectrophotometrically. The selection of the type of resin was critical, the criteria being that the paraquat, but not the interfering species, adsorb well onto the column, that much of the light-absorbing material retained by the resin be eluted before the paraquat, and that the latter be eluted with high efficiency. The procedure was found useful for waters containing substances absorbing light in the region of 256 mμ, the absorption maximum for unreduced paraquat. For water samples of 50 to 500 ml the expected recovery of the method is 85 to 100%.

Finally, it is of interest to note an application for the removal of interferences in the low level measurement of oxygen in boiler waters.[36] Such waters frequently contain iron(II) and hydrazine, the latter being added as a scavenger for dissolved oxygen. In the Winkler analysis of these waters containing oxygen in the range of 0.01 to 0.04 mg/l, hydrazine and iron(II) were found to interfere when in the range of 0.04 to 0.2 mg/l and 0.1 to 1.2 mg/l respectively. By first passing the test solution through a cation exchange resin, these interferences were successfully removed.

Chromatography

In many cases the most difficult part of an organic analysis is the separation of mixed components into relatively pure fractions prior to analytical operations. Within the last fifteen years several separation techniques have proven so successful that they serve today not only as separation tools but as sensitive methods of analysis *per se*. The various separation or "chromatographic" techniques are categorized primarily on an operational rather than a conceptual basis. Thus, the terms, paper, column, thin-layer, gas-liquid, and ion exchange chromatography, for example, refer to different procedural applications of the same fundamental concept.

General principles

A rigorous presentation of chromatographic theory is beyond the scope of this manual. Adequate treatments of theory are available in the literature,[1,9,13,31] although such presentations are often confined to a specific chromatographic technique.

All forms of chromatography consist of at least two immiscible phases, one static and the other mobile. The static phase may be a solid or a liquid held on a solid, whereas the mobile phase may be a gas, a liquid or a dissolved solid. In all cases the mobile phase either is or contains the sample. Types of chromatography may be classified according to the nature of these phases in any particular combination as shown in Table 9.

Separation of components is due to phase equilibria that occur between the sample components and the static mobile phases. This results in a distribution or "partitioning" of the sample between the two phases. At equilibrium a solute will distribute itself between two immiscible phases so that its chemical potential (μ) or escaping tendency is equal. Therefore,

$$\mu_1 = \mu_1{}^* + RT \ln C_1 \quad \text{(static phase)} \quad (9)$$

$$\mu_2 = \mu_2{}^* + RT \ln C_2 \quad \text{(mobile phase)} \quad (10)$$

Table 9

Classification of Chromatographic Methods*

Moving Phase	Phase	Name of Method
L	S	column chromatography thin layer chromatography ion exchange chromatography≠
L	L	column chromatography paper chromatography
G	S	gas-solid chromatography molecular sieves
G	L	gas-liquid chromatography

* L = Liquid; G = Gas; S = Solid

≠ Mobile phase is dissolved solid

and at equilibrium

$$\mu_1 = \mu_2 \tag{11}$$

$$\ln \frac{C_1}{C_2} = \frac{\mu_1 - \mu_2*}{RT} \tag{12}$$

$$\frac{C_1}{C_2} = \exp\left(\frac{\mu_1* - \mu_2*}{RT}\right) = P \tag{13}$$

The expositional term is independent of concentration and is known as the "partition coefficient" (P). Equation 13 is essentially the same as Equation 3. Each particular solute-static phase-mobile phase combination will have a characteristic coefficient that will determine the migration rate of each solute through the system at any given temperature and flow rate of mobile phase. Regardless of the nature of the force responsible for partitioning (i.e., solubility, adsorption, chemical bonding), the intensity of the force will vary among the components of a sample. Therefore, as contact between the phases increases, the separation of the components will increase.

Gas-Liquid chromatography

In this form of chromatography volatile materials are separated by passing a gas stream over a finely divided solid static phase which may be coated with a high boiling liquid. Many variations in apparati are (commercially) available in varying degrees of sophistication and cost ($1000 - $15,000). A basic block diagram of a typical instrument is shown in Figure 3. A chromatogram is obtained as a record on a strip chart recorder which continuously receives a signal from a detector monitoring some physical or chemical property of the effluent stream. Any material having an appreciable vapor pressure (1-1000 mm) at the temperature of operation (0-400°C) can be satisfactily eluted.

Separation Process

The partitioning of sample components in the column is subject to plate theory analysis.[31] Employing this theory as a model it is possible to identify operational characteristics required for ideal chromatographic separation. Gas flow through the column must be constant and no axial diffusion permitted in any phase; *i.e.*, components must enter and leave the two phases simultaneously, thus providing instantaneous equilibrium. For any column of given composition and geometry, an optimum gas flow will exist for ideal separation of two sample components. At flows less than optimum, longitudinal diffusion occurs and column efficiency is reduced. For ordinary columns (5-6 mm I.D.) 20 ml/min is sufficient to avoid this problem. At high flows insufficient time is allowed for equilibration. Column efficiency is related to the number of theoretical plates calculable from plate theory,[1] and is largest when the ratio of the volume of carrier flow needed for elution of a component to the width of the peak is large.

Columns

Chromatographic columns are generally of three types:

(A) analytical: liquid coated or solid support, usually restricted to 20 ft. lengths due to pressure drop; 1/4" to 1/8" O.D., 0.05 - 0.01" I.D.

(B) capillary: liquid coated on inner wall of tubing; 100-200 ft.; 1/16" O.D., 0.01" I.D.

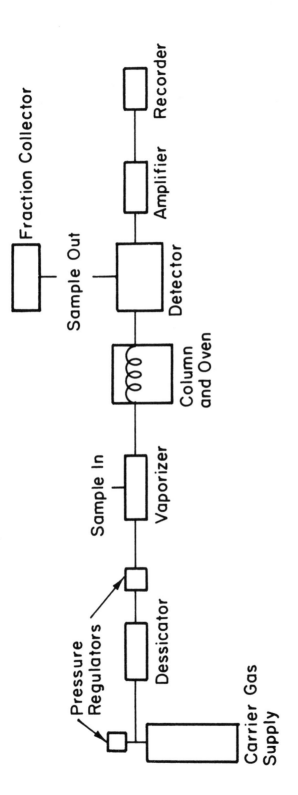

Figure 3. Block diagram of typical gas chromatograph

(C) preparative: about 20 ft. maximum length;
usually 3/8" O.D.

All columns are constructed of stainless steel, glass
or aluminum, with stainless steel being most popular
because of its cost and durability. The columns are often
shaped in coils because of their great lengths, and in
such cases the coil diameter should be at least ten times
the column diameter to avoid diffusion effects.

Many columns are packed with solids used primarily as
support for active liquid coatings. Solid supports should
be inert, have large surface areas and be of uniform size.
Basic support materials are:

(A) Chromosorb P: crushed firebrick; shows some
adsorption effects; used for high
efficiency packed columns.
Surface area about 4 m^2/g.

(B) Chromosorb W: prepared from filter aids; inert;
less efficient than Chromosorb
P. Surface area about 1 m^2/g.

Liquid phases used to coat solid supports are usually
high boiling polymers, such as ethylene oxide polymers,
high molecular weight alcohols, oils and greases. Proper
selection of the liquid phase is essential to successful
chromatography, and references are available[29] which
describe the types of liquid coatings needed for a variety
of specific applications. Liquid phases are coated on
solids in the 2-10 wt% range; lower loadings require very
inert solid support and higher loadings give reduced
efficiency. Vapor pressures of liquid coatings must be
low (<0.01 mm) at operating temperatures to prevent
"bleeding" during operation. Bleeding not only shortens
column life but reduces efficiency and interferes with
detection. Therefore each liquid coating has a maximum
recommended temperature for operation.

Columns may be obtained commercially and assembled to
specification; however, this procedure is often expensive.
Stock quantities of solid support and pure liquid phases
in Kg quantities may be purchased at reasonable prices and
the coating performed in the laboratory. The liquid phase
is dissolved in an appropriate solvent and mixed with the
solid support in either a rotary evaporator or a saucepan
with shaking. New columns must be conditioned by running
overnight at 20°C above normal operating temperature with
small carrier gas flow (5-10 ml/min).

Gas flow rates through columns during operation vary
widely, but are generally in the 20 to 60 ml/min range

for 1/8" columns and in the 80 to 200 ml/min range for 1/4" columns.

Sample Introduction

Liquid and gas samples may be injected with a calibrated syringe through a septum onto the head end of the column. Injector blocks on most instruments are maintained 8-10°C warmer than the oven to flash evaporate liquid samples and prevent condensation. Sample volumes must be as small as possible and injection performed as swiftly as possible to avoid nonideal effects. Since the injector block is generally at 5 to perhaps 60 psi pressure, extreme care must be given to injection technique or nonreproducible data will result.

Carrier Gas

The nature of the inert gas used to carry the sample through the column depends primarily on the analysis and the type of detector. Helium, argon, hydrogen and nitrogen are frequently used, although any gas that is distinguishable from sample components may be used. The use of impure carrier gas may cause extraneous peaks on the chromatogram or baseline drift of the recorder.

Commercial grade gases may be dried by passing them through a molecular sieve trap which can be reconditioned by heating to 400°C for four hours with a gas stream passing through it. A convenient check on the impurities in a carrier gas supply can be performed by withdrawing a syringe load of carrier and reinjecting it at operating conditions. Impurities will then enter the column as a plug flow and appear in the detector in addition to background impurities.

Detectors

Ideally, detectors give one signal when carrier gas is present and a different signal when carrier gas plus a sample component is present. (The signal produced by the carrier gas alone is adjusted to the recorder baseline or zero.)

Detectors vary widely in the mechanism of their response and in their sensitivities and specificities. It is generally true that as the sensitivity of detectors increases, the range of compounds they are capable of responding to decreases. Thus the flame ionization detector is much more sensitive than the thermal conductivity detector, but the former responds only to organic molecules whereas the latter responds to any molecule different from the carrier gas.

Detectors are often characterized by the minimum detectable quantity that results in signal production (MDQ), the concentration range over which the detector responds (Dynamic Range, DR), and the concentration range over which the detector response is linear (LDR).

(A) Thermal conductivity detection

A heated platinum filament will lose heat mainly by conduction through surrounding gas. The heat conductivity of the gas depends on the mobility of the gas molecules which is a function of molecular weight. Thus if the composition of a gas stream passing over heated Pt filaments changes (due to elution of sample component) the filament temperature will change. The filaments in commercial chromatographs are made part of a Wheatstone Bridge so that resistance changes due to the temperature effects can be measured, amplified and recorded. Thermal conductivity (TC) filaments are usually coiled and possess a high temperature coefficient of resistance. The wires are approximately 100°C warmer than surrounding parts of the detector which are at approximately 300°C.

MDQ	2-5 micrograms
Response	all components except carrier
LDR	10,000
Carrier Gas	He
Temp. Limit	450°

(B) Flame ionization detection

Carrier gas and sample mixtures leaving the column are burned in an H_2-O_2 flame at a quartz tipped jet.

$$H_2 + O_2 + \text{Organic} \rightarrow CO_2 + H_2O + e^- + \text{ions}$$

$$\Sigma\ e^- + \text{ions} \rightarrow \text{current}$$

An ion collector electrode is situated directly above the flame and is maintained at a potential of 200-300 volts. Current flows from the collector electrode to an electrometer.

MDQ	10^{-11}g
Response	All organic compounds. Insensitive to all fixed gases, H_2S, CO_2, SO_2, CO, H_2O, etc.
LDR	10^6
Carrier Gas	N_2, He, Argon
Temp. Limit	400°C

56

(C) Electron capture detection

The electron capture detector employes a tritium source (titanium tritide on steel foil) to ionize nitrogen carrier gas molecules producing slow electrons. These slow electrons migrate to the anode under a fixed voltage producing a steady current which is amplified by the electrometer. If a sample contains electron-absorbing materials, this current will be reduced. The loss of current is a measure of the amount and electron affinity of the component.

$$\beta + N_2 \rightarrow N_2^+ + e^-_{slow}$$

$$e^-_{slow} + X \rightarrow X^-$$

$$\text{Loss of } e^- \rightarrow \text{Reduces current}$$

MDQ	Variable, sensitive to halides, metal organics, S-compounds
LDR	500
Carrier Gas	N_2 (very pure)
Temp. Limit	$220^{\circ}C$

(D) Microcoulometric detection

This process detects components that will produce unionized compounds with Ag^+, *e.g.*, sulfur and the halogens. The detector consists of an electrolytic cell containing silver electrodes immersed in silver acetate. When a component reacts with Ag^+ by dissolution of the Ag electrode, the current flow is amplified and recorded.

Data Interpretation

Information obtained directly from the chromatogram is of both qualitative and quantitative significance. Frequently used parameters are:

    (A) retention volume: volume of carrier gas required to elute component from injection to detection.

    (B) retention time: calculable from retention volume under conditions of constant flow; measurable

	with stopwatch from injection
	to detection.
(C) retention	distance on recorder chart
distance:	under constant flow and con-
	stant chart speed, from
	injection to detection.
(D) adjusted	measured from air or solvent
retention volume:	peak to detection of com-
(time or	ponent. Eliminates column
distance)	dead volume.
(E) relative	measured relative to standard
retention volume:	component appearing in same
(time or	chromatogram.
distance)	

Relative retention data (E) are preferable as they are subject only to temperature variations in the column during development, whereas the other parameters are subject to flow variations as well. Qualitative identifications are made by comparison with known compounds on several columns. Quantitative measurements are made by correlating peak height (or area for asymmetrical peaks) with concentration of standards.

Other data are often valuable for compound identification. The comparison of response ratios of a given compound analyzed by two different detectors under fixed conditions is characteristic of that compound. In addition, nonchromatographic identification of separated and collected components by derivative formation, spectrophotometric measurement, or mass spectrometry is possible.

Thin-Layer Chromatography

Thin-layer chromatography is analytical adsorption chromatography. It involves spreading a thin layer of adsorbent on a solid flat support such as a glass plate, applying a mixture of components to a spot on one end of the adsorbent, and developing the chromatogram by immersing the tip of the coated plate in a suitable solvent. The procedure is fast and has good sensitivity for the lipophilic materials difficult to separate by paper chromatography. Excellent reviews[37] are available detailing specific adsorbent-solvent combinations for given compound types (see Chapter X).

Separated compounds may be fluorescent under UV examination of the plate or may be located with various chromogenic spray reagents. The ratio of the distance traveled by any component to that traveled by the solvent front is characteristic of a given compound. Internal standards are helpful as in gas-liquid chromatography.

Other Techniques

Although selective precipitation has been widely used in the gravimetric analysis of nontrace elements, co-precipitation is of particular interest as a concentration technique for improving sensitivity in the analysis of trace species. For reviews and applications of co-precipitation in water analysis the reader is referred to articles by Riley[38] and Joyner et al.[21] Co-precipitation of many metals in sea water occurs with ferric hydroxide as the primary precipitate. In many cases close to 100 per cent recovery may be obtained, and the metals may be analyzed in the precipitate itself. Neutron activation analysis, for example, or analysis after subsequent re-solution may be used. Organic complexes have also been used to precipitate many trace elements in natural waters. In one such application 17 trace elements were analyzed by DC arc emission spectroscopy after ashing the precipitate.[44]

Various membrane processes have been utilized for concentration and separation purposes in water analysis. Thus, membrane dialysis has been used to separate and demonstrate the presence of high molecular weight material in tap water.[7] As was noted earlier, suspended and particulate matter may be separated from soluble species and most colloidal matter by membrane filtration. This technique is widely utilized and a portable field apparatus for such a purpose has been recently described.[45]

Distillation, evaporation, and sublimation have been used to concentrate or separate in the analysis of aqueous solutions. Thus, freeze drying has been found to be a useful technique for concentrating trace elements in sea water[42] and soluble organic matter in fresh water[32] prior to analysis. Similarly, heat evaporation is widely used for this purpose, but has some limitations due to volatilization, decomposition, or oxidation of organic or inorganic constituents.[47] However, vacuum rotary evaporation at temperatures of 40-50°C has been shown to give higher recoveries in shorter times. Distillation is frequently used to eliminate or reduce interferences in many standardized methods of analysis, including those for ammonia, phenol, and fluoride.

Centrifugation may be used to separate various species in water. Thus, continuous-flow and density-gradient techniques have been used to separate suspended and colloidal particles, macromolecules, viruses, and bacteria.[4,27] Although electrophoresis has not been widely used in water analysis, an instrument for continuous particule electrophoresis is available for separating and collecting colloidal species and particles as large as 100 microns.[8] Adsorptive bubble techniques, such as foam fractionation, have also not received wide attention as a concentration or separation method in water analysis, but have some potential for this purpose. For reviews the reader is referred to articles by Rubin[39] and Lemlich.[28]

Although these various separation and concentration techniques are useful and meet most needs of the analyst, they do introduce another step into the analytical process. Thus, wherever possible, interferences should be removed by masking techniques, or a sufficiently sensitive analytical procedure chosen so as to eliminate the need to concentrate.

REFERENCES

1. American Association of Professors in Sanitary Engineering (AAPSE), "Fundamentals of Chromatography," Second Annual Workshop, University of Colorado, June, 1967.
2. American Public Health Association, "Standard Methods for the Examination of Water and Wastewater," 12th edition (New York:1965) p 213.
3. Andelman, J.B., and S.C. Caruso. *Handbook of Water and Water Pollution*, L.L. Ciaccio, Ed., Vol. 1 (New York: Marcel Dekker, in press, 1965) Chapter 13, p 213.
4. Anderson, N. G., G. B. Cline, W. W. Harris, and J. G. Green. "Transmission of Viruses by the Water Route," G. Berg, Ed. (New York: Wiley-Interscience, 1967) pp 75-88.
5. Baker, R. A. "Microorganic Matter in Water," ASTM STP 448 (Philadelphia: American Society for Testing and Materials, 1968) p 65.
6. Baker, R. A. J. Water Pollution Control Federation 37, 1164 (1965).
7. Barth, E. F., and N. H. Acheson. J. Am. Water Works Assoc. 54, 959 (1962).
8. Beckman Instruments, Inc., "CPE System for Continuous Particle Electrophoresis," Bulletin 7096-B, Fullerton, Calif. (1967).
9. Berg, E. W. *Physical and Chemical Methods of Separation* (New York: McGraw-Hill, 1963).
10. Biechler, D. G. Anal. Chem. 37 1054 (1965).
11. Boyd, G. E., A. W. Adamson, and L. S. Myers, Jr. J. Am. Chem. Soc. 69, 2836 (1947).
12. Calderbank, A., and S. H. Yuen. Analyst 90, 99 (1965).

13. Cassidy, H. G. *Techniques of Organic Chemistry*, A. Weissberger, Ed., Vol X (New York: Interscience, 1957).
14. Christman, R. F. "Chemical Structures of Color Producing Organic Substances in Water," Symposium on Organic Matter in Natural Waters, Institute of Marine Science, University of Alaska, 1968.
15. Craig, L. C., and O. Post. Anal. Chem. 21, 500 (1949).
16. Fishman, M. J., and M. R. Midgett. *Trace Inorganics in Water*, R. F. Gould, Ed., Advances in Chem. Series No. 73, (Washington, D.C.: American Chemical Society, 1968) Chapter 12.
17. Greenhalgh, R., J. P. Riley, and M. Tongudai. Anal. Chim. Acta 36, 439 (1966).
18. Helfferich, F. *Ion Exchange* (New York: Mc-Graw-Hill, 1962).
19. Hoak, R. D. Intern. J. Air Water Pollution 6, 521 (1962).
20. Inczedy, J. *Analytical Applications of Ion Exchangers* (New York: Pergamon Press, 1966).
21. Joyner, T., M. L. Healy, D. Chakravarti, and T. Koyanagi. Env. Sci Tech. 1, 417 (1967).
22. Kawahara, F. K., J. W. Eichelberger, B. H. Reid, and H. Stierli. J. Water Pollution Control Federation, 39 572 (1967).
23. Kelso, F. S., J. M. Mathews, and H. P. Kramer. Anal. Chem. 36, 577 (1964).
24. Kobayashi, S., G. F. Lee. Anal. Chem. 36, 2197 (1969).
25. Kolfenbach, J. J., E. R. Kooi, E. I. Fulmer, and L. A. Underkoffer. Ind. Eng. Chem., Anal. Ed. 16, 473 (1944).
26. Laitinen, H. A. *Chemical Analysis* (New York: McGraw-Hill, 1960).
27. Lammers, W. T. Env. Sci. Tech. 1, 52 (1967).
28. Lemlich, R. Ind. Eng. Chem. 60 (10) 17 (1968).
29. Lynn, T. R., C. L. Hoffman, and M. M. Austin. *Guide to Stationary Phases for Gas Chromatography* (Hamden, Connecticut: Analabs, Inc., 1968).
30. McCoy, J. W. Anal. Chim. Acta 6, 259 (1952).
31. McNair, H. M. and E. J. Boneili. *Basic Gas Chromatography* (2700 Mitchell Drive, Walnut Creek, California: Varian Aerograph, 1967).
32. Midwood, R. B. and G. T. Felbeck, Jr. J. Am. Water Works Assoc. 54, 959 (1962).
33. Navone, R. J. Am. Water Works Assoc. 46, 449 (1954).
34. Nydahl, F. Proc. Intern. Assoc. Theor. Applied Limnology 11, 276 (1951).

35. Platte, J. A. *Trace Inorganics in Water*, R. F. Gould, Ed., Advances in Chem. Series No. 73 (Washington, D.C.: American Chemical Society, 1968) Chapter 14.

36. Potter, E. C. J. Appl. Chem., **9**, 645 (1959).

37. Randerath, K. Thin Layer Chromatography (New York: Verlag Chemie, Academic Press, 1963).

38. Riley, J. P. *Chemical Oceanography*, J. P. Riley and G. Skirrow, Eds., Vol. 2 (New York: Academic Press, 1965) Chapter 21.

39. Rubin, A. J. J. Am. Water Works Assoc. **60**, 832 (1968).

40. Samuelson, O. Z. Anal. Chem. **116**, 329 (1939).

41. Samuelson, O. *Ion Exchange Separation in Analytical Chemistry* (New York: Wiley, 1963).

42. Schutz, D. F., and K. K. Turekian. Geochim. Cosmochim. Acta **29**, 259 (1965).

43. Shapiro, J. Science **133**, 2063 (1961).

44. Silvey, W. D., and R. Brennan. Anal. Chem. **34**, 784 (1962).

45. Skougstad, M. W., and G. F. Scarbro, Jr. Env. Sci. Tech. **2**, 298 (1968).

46. Skrinde, R. T., and H. D. Tomlinson. J. Water Pollution Control Federation, **35**, 1292 (1963).

47. Slonim, A. R. and F. F. Crawley. J. Water Pollution Control Federation **38**, 1609 (1966).

48. Stephens, B. G., and H. A. Suddeth. Anal. Chem. **39**, 1478 (1967).

49. U. S. Public Health Service, "Drinking Water Standards," Public Health Service Pub. 956, 1963.

50. U. S. Public Health Service, "Summary Report, Advanced Waste Treatment Research," AWTR-14, Washington, D.C., 1965.

Chapter V

Elements of
Instrumental Analysis

This chapter reviews briefly the principles of
selected instrumental methods of chemical analysis. These
methods have been used to varying degrees for water quality
characterization for pollution control. It must be
realized that no attempt has been made to provide an
exhaustive coverage of the subject matter. For more complete coverage the reader is referred to textbooks in
analytical chemistry or instrumental analysis.

Chemical Spectrophotometry - General

As the name implies, the technique utilizes the absorption of portions of the electromagnetic spectrum. Figure
4 illustrates the various portions of the electromagnetic
spectrum. The figure also serves to indicate the general
areas of the spectrum in spite of the fact that the
boundaries between the various zones are somewhat
arbitrary.

The energy in an atom is solely electronic and exists
in discrete quantized energy levels. Accordingly, atomic
emission or absorption spectra consist of sharp lines of
specific spectral wave lengths. On the other hand, the
energy of polyatomic molecules consists of electronic
energy, which is involved with the electrons in the atoms,
rotational energy, which is involved with the rotation of
a molecule, and vibrations of the atoms relative to each
other along their internuclear axes. These energies are
also quantized, and absorption or emission occurs at
specific wave lengths. Nevertheless, these energies are
interdependent, so when one type of energy is affected,

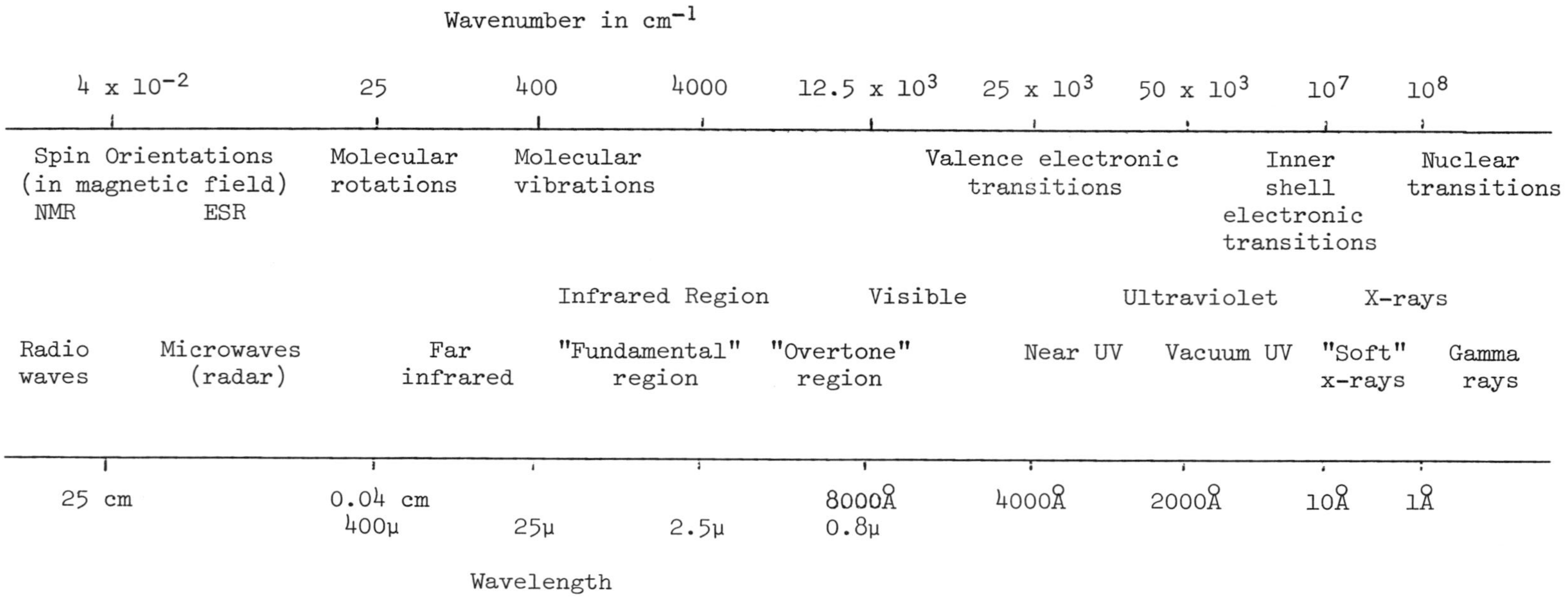

Wavenumber in cm^{-1}

4×10^{-2}	25	400	4000	12.5×10^3	25×10^3	50×10^3	10^7	10^8

Spin Orientations (in magnetic field) NMR ESR	Molecular rotations	Molecular vibrations		Valence electronic transitions		Inner shell electronic transitions		Nuclear transitions
		Infrared Region		Visible		Ultraviolet		X-rays
		Far infrared	"Fundamental" region	"Overtone" region	Near UV	Vacuum UV	"Soft" x-rays	Gamma rays
Radio waves	Microwaves (radar)							

25 cm	0.04 cm 400μ	25μ	2.5μ	8000Å 0.8μ	4000Å	2000Å	10Å	1Å

Wavelength

Figure 4. The electromagnetic spectrum.

the other two are also affected. This results in spectra
with broad bands at the specific wave lengths instead of
discrete lines.

Absorption in the ultraviolet and visible regions is
mainly electronic in nature and is associated with resonating structures in the molecules. In the infrared
region, the absorption is due to the vibrational energies
of the groupings in the molecule and the rotational
energies of the molecule itself. By observing the absorption characteristics of a chemical material, it is possible
to gain information concerning the qualitative nature and
quantitative composition of that material in a given test
solution.

The optical phenomena and techniques principally discussed in this chapter are:

1. the absorption of radiation by dissolved
    molecules: molecular absorption spectrophotometry,
2. the absorption-remission of radiation by dissolved
    molecules: molecular fluorescence spectrophotom-
    etry,
3. the absoprtion of radiation by free atoms:
    atomic absorption spectrophotometry, and
4. the absorption-remission of radiation by free
    atoms: atomic fluorescence spectrophotometry.

Molecular Absorption Spectrophotometry

This technique is based on the absorption of radiation
by molecular species. It has been used for the analysis
of metal ions in natural and waste waters and is based
primarily on reacting metal ions with various organic
reagents to form colored compounds which may be determined
spectrophotometrically either directly or after appropriate
separation. A complexometric reaction between the metal
ion and the organic molecule--acting often as a multi-
dentate ligand--is usually involved.

Although very many organic compounds absorb quite
strongly, only a limited number of inorganic ions do, and
it is the normal procedure of inorganic absorption spectrophotometry to add a molecule or reagent species to the
solution of the inorganic ion which will react with it
and, in the process, bring about a marked change in the
spectral absorption characteristics of the reagent. It is
necessary that the absorption spectrum of the reagent-ion
be well-separated in at least one place from the absorption
spectrum of the reagent itself. It is common practice to
add a fairly large excess of the organic reagent so that
virtually all of the ionic species are driven to react with
the reagent.

65

Some of the more common organic reagents used for separation by extraction are chelate compounds, *e.g.*, dithizone (diphenylthiocarbazone), oxines, cupferron, and diethyldithiocarbamate. Examples of the applications of this technique for analysis of metals in natural and waste waters are given in Table 10. The basic law of absorption spectrophotometry relates the absorbance "A" to the concentration of the absorbing species "C" linearly through the length of the absorbing layer of solution "l" and the molar absorptivity of the absorbing species, "k,"

$$A = \log (I_0/I) = klC \qquad (14)$$

where I_0 and I are the intensities of the incident and emitted light respectively. Accordingly, it seems that the analytical signal can be increased for a given concentration of the reacting ion C by increasing the path length l or by using a compound which has a greater molar absorptivity k. This is true for small changes in l, but the background absorption by other agents in solution becomes limiting for very large light path lengths. Increasing the sensitivity of such determinations ultimately depends on the formation of color compounds of high molar absorptivity.

Sensitivity limits on molecular absorption spectrophotometry for metal analysis are about 10^{-6}M. It is possible, however, to extend this sensitivity by using differential spectrophotometric techniques which also allow for more precise determinations than are possible by conventional procedures. A schematic diagram of a double beam spectrophotometer is shown in Figure 5.

In this type of instrument arrangement, one beam provides a reference signal, the other, a measuring signal. The blank is placed in the first beam, the sample, in the second. By referring intermittently or continuously to the reference beam when making measurements, errors that arise from fluctuations and drift in the applied voltage, source intensity, detector response, amplifier gain, and other irregularities, as well as errors that the blank is particularly designed to compensate for, are largely eliminated. These are the main advantages of double beam spectrophotometers over single beam instruments.

The generalized diagram of the double beam spectrophotometer shown in Figure 5 has an automatic optical null balance. The reference and sample beam signals reach the detector alternately. If they are of equal intensity, the AC amplifier has no output of "unbalance" signal, hence no power is fed to the servomotor driving the optical wedge. If the intensities of the two beams are different, an

Table 10

Examples of Molecular Absorption Spectrophotometry
for Metals in Natural and Waste Waters

Metal	Complexing Agent	Solvent Extraction	Color of Complex	pH Range	Suitable Wave Length mμ	Useful Range mg/1
Cobalt	diethyldithio-carbamate	ethylacetate	blue	acidic pH 3.0	367	---
Cadmium	dithizone	carbon tetrachloride	red	alkaline pH 10-12	518	0.1-5
Chromium	1,5 diphenyl carbohydrazide	butanol	violet	acidic pH 2-3	540	0.05-0.5
Copper	dithizone	carbon tetrachloride	violet	acidic pH 0.5	510	0.04-14
Copper	diethyldithio-carbamate	carbon tetrachloride	yellow-brown	alkaline pH 9.0	436	0.1-0.8
Copper	cuprione (2-2'diquinolyl)	isoamyl alchohol	purple	pH 5-6	540	---

Table 10
(cont)

Metal	Complexing Agent	Solvent Extraction	Color of Complex	pH Range	Suitable Wave Length mμ	Useful Range mg/1
Iron	O-phenanthroline	---	orange-red	pH 209	490	0.01-1.0
Iron	thioglycollic acid	---	purple	pH 8-12	540	0.04-1.2
Iron	tripyridyl	---	red-purple	pH 9-10	560	0.01-2.0
Lead	dithizone	chloroform	red	pH 7-10	520	---
Mercury	dithizone	carbon tetrachloride	yellow-orange	pH 0-1	500	---
Nickle	dimethylyoxime	---	reddish-brown	neutral	465	---
Zinc	dithizone	chloroform	purple-red	pH 4-5.5	530	0.1-1.0
Zinc	zincon	---	blue	pH 9.0	620	0.1-2.4

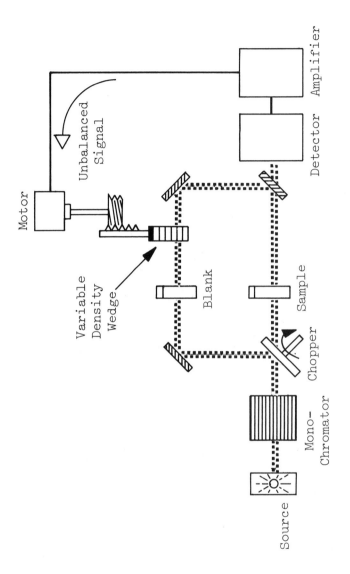

Figure 5. Double beam-null balance spectrophotometer.

"unbalance" signal is generated which drives the wedge, causing an increase or decrease in the reference signal. This process continues until the intensity of the two beams impinging on the detector are equalized again. Recording spectrophotometers contain other motors for wave length scanning and automatic recording.

Molecular Fluorescence Spectrophotometry

Molecular fluorescence spectrophotometry is based on the spectral measurements of fluorescence radiation emitted from luminescent compounds upon excitation by incident radiation. The reemitted radiation is of lower frequency than the absorbed light. Fluorescence spectra are characteristic of the compound in the sense that the emission spectrum is always the same irrespective of the wave length of the incident light which promotes fluorescence.

The fluorescence equation may be expressed as follows

$$F = 2.303 \; \phi \; I_0 k \ell p \; C \qquad (15)$$

where F is the amount of fluorescence generated, ϕ is a constant related to the efficiency of fluorescence, I_0 is the intensity of incident radiation, k is the molar absorptivity at a given wave length, ℓ is the path length in centimeters, p is a fractional constant, and C is the concentration. Accordingly, F measured in terms of signal response of a photo-multiplier tube sensitive to fluorescence radiation is proportional to the analytical concentration C, while the parameters I_0, ℓ and p are instrumental factors, and parameters ϕ and k are functions of the efficiency of the fluorescent reagent system.

In the case of absorption spectrophotometry, Equation 14, any increase in I_0 will be accompanied by a matching increase in I, with no net gain in the absorbance, A. But, in the case of fluorescence spectrophotometry, any increase in I_0 will be matched by a corresponding increase in the analytical signal F, as indicated in Equation 15. It is interesting to note also that any increase in the amplifier gain in absorption spectrophotometry will amplify both I_0 and I correspondingly, whereas in fluorescence spectrophotometry this will result only in an increase in F. Accordingly, the sensitivity of molecular fluorescence spectrophotometry is inherently greater than that of molecular absorption spectrophotometry. The technique is extremely useful and is applicable to solutions 100-10,000 times more dilute than those which can be analyzed by absorption spectrophotometry.

A typical instrument used for analytical applications of

70

fluorescence phenomena is shown in Figure 6. The essential parts are (a) a detector to sense the level of the admitted radiation, usually with an amplifier to enhance the analytical signal, (b) a monochromator or filter system to disperse the fluorescent radiation before it reaches a detector, (c) a cell or cuvette to hold the solution, (d) a monochromator or filter system to disperse the excitation radiation, and (e) a source of radiation of suitable intensity and emission characteristics. Fluorescence usually is observed at right angles to the instant beam of light in order to eliminate contamination of fluorescence signals by radiation from the excitation wave length.

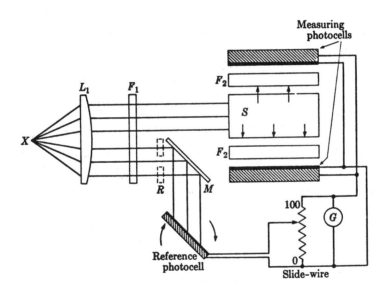

Figure 6. A simplified schematic diagram of a filter fluorometer, the Lumetron Model 402-EF. L_1, collimating lens; F_1, primary filter passing only UV; F_2, secondary filter passing only fluorescent light; R, reduction plate; M, front-surface mirror; G, galvanometer.

There are a number of fluorometric reagents suitable for analysis of such metals as aluminum, rare earths, zinc,

and calcium. Measurements are based on the extinction of
the fluorescence of the reagents with which they react.
A typical example of an effective fluorogenic agent is
8-hydroxyquinoline, which forms fluorescent complexes with
aluminum, beryllium, etc. and nonfluorescent complexes
with iron, copper, etc. Perhaps one of the most desirable
characteristics of molecular fluorescence spectrophotometry for the analysis of natural and waste waters is its
selectivity. There are only certain ions which are
capable of producing fluorescence. While over 30 metal
ions produce absorption spectra, only few are capable of
producing fluorescence with a nonselective agent like
8-hydroxyquinoline.

Certain phenomena have deleterious effects on the
efficiency of fluorescence of molecular species and may
result in the extinction of fluorescence. This is mani-
fested by a decrease in the quantum efficiency factor, ϕ,
which relates the number of quanta emitted to the number
of quanta absorbed. Ideally, ϕ should have a value of
unity and under favorable circumstances compounds like
Rhodamine B and fluorescein have been reported[45] to
exhibit high quantum efficiencies approaching unity.

The reduction of the quantum efficiency factor ϕ[8] can be
a result of (a) collisional deactivation of excited state
molecules with others, which explains the marked decrease
of fluorescence with increase in temperature, (b) the
nature of the solvent, *e.g.*, strong polar solvents have an
adverse effect on fluorescence relative to nonpolar solvents, (c) the presence of anions or cations which may
interact with the fluorescence molecules, and (d) the
presence of dissolved oxygen, which seems to have a
quenching effect on many fluorescent organic molecules.

Molecular fluorescence spectrophotometry has been used
for the analysis of trace organic matter in natural waters
and lignin sulfonate compounds in spent sulfite effluents
from Kraft processes.[11] Generally, aliphatic compounds
rarely exhibit fluorescence because they absorb weakly in
the near UV or visible range and hence cannot be expected
to fluoresce. Aromatic compounds, on the other hand,
possess good π bonding systems and are commonly fluorescent. Molecular configuration and steric hindrance
effects have been reported to give recognizable patterns
in molecular fluorescence.[8] A good variety of fluorometric reagents is available for trace inorganic
analysis.[47] Table 11 presents selected methods for trace
inorganic ions for natural and waste waters characterizations.

Table 11

Spectrofluorometric Methods for Certain Inorganic Ions

Determination	Reagent	Conditions	Maxima		Sensitivity	Interferences
			Absorption mμ	Fluorescence mμ		
Al	Alizarin Garnet R(CI 168)	pH 4.6	470	500	0.007 μg/ml	Be,Co,Cr,Cu,F$^-$, Fe,NO$_3^-$,Ni,PO$_4^{-3}$, Th,Zr
Mg	Bissalicylidene diaminobenzofuran	pH 10.5	475	545	0.002 μg/ml	Mn
Mo	Carminic acid	pH 5.2	560	590	0.1 μg/ml	
Cd	2-(o-hydroxyphenyl) benzoxazole	pptd., dissolved in HOAc	365	Blue	2 μg/ml	NH$_3$
Cu(II)	Tetrachlorotetraio- dofluorescein o-phenanthroline	pH ≅ 7	560	570	0.001 μg/ml	Cyanide

Table 11
(cont)

Determination	Reagent	Conditions	Maxima		Sensitivity	Interferences
			Absorption mμ	Fluorescence mμ		
F⁻	Al complex of alizarin garnet R or Eriochrome red B	pH 4.6 (quenched reaction)	470	590	0.001 μg/ml	Be,Co,Cr,Cu, Ni,PO_4^{-3},Tl
Be	1-amino-4-hydroxyanthraquinone	0.02 M NaOH	540	620	0.2 μg/ml	CrO_4,Li
	2,3-hydroxynaph-thoic acid	pH 7.5	380	460	0.0002 μg/ml	Sc,Cr(III), Bi(III),Ce, Sn(IV),Th(IV), Fe
	8-hydroxyquin-aldine	(CHCl₃)pH 8.0	–	–	0.001 μg/ml	Al,Bi,Cd,Cr, Cu,Fe,In,Sn, Ti,Zn

Atomic Absorption Spectrophotometry

Atomic absorption spectrophotometry is a relatively new technique which is gaining great popularity in the analysis of natural and waste waters. The technique is really a combination of emission and absorption phenomena, and closely resembles flame photometry. In flame photometry, a flame excites the elements in the sample to produce an emission spectrum. However, only a small percentage of the atoms are excited. Atomic absorption increases the sensitivity of the flame technique by utilizing the unexcited atoms in the flame. In atomic absorption, as in flame photometry, the sample solution is atomized into a flame, producing atomic vapor of the elements in question. A monochromatic light from a hollow cathode tube containing the desired element and emitting light of the same wave lengths as that of the desired element is passed through the atomic vapor of the sample in the flame. The atoms of the desired element in the vapor are mainly in their unexcited or ground state in the flame, and they absorb the radiation from the light source. The amount of light absorbed is proportional to the amount of the element in the sample.

The similarity of atomic absorption spectrophotometry to molecular absorption spectrophotometry is based on the fact that atoms are capable of absorbing light in exactly the same way as molecules by interacting with the photons of energy required to promote an electronic transition from ground state to one of the excited states of the atom. Accordingly the laws which govern the relationship between the amount of light absorbed and the concentration of the absorbing species, as well as the experimental apparatus and techniques, are basically the same for both atomic and molecular absorption spectrophotometry.

Figure 7 shows a schematic diagram of an atomic absorption spectrophotometer. In this single beam instrument, a mechanical modulated chopper alternately passes and reflects the light beam, thus creating two separate equal beams. One beam by-passes the sample, and its intensity is measured as I_0. The second beam passes into the sample, and its intensity is measured as I. The absorbed light is determined as $I_0 - I_1$. Corrections for any absorption by the flame and combustion products of the sample must be made. As an analytical tool, atomic absorption spectrophotometry has the unique advantage of virtual specificity. Exceptions are those few cases in which unfavorable matrix components are present in the sample solution. This is, to a great extent, the result

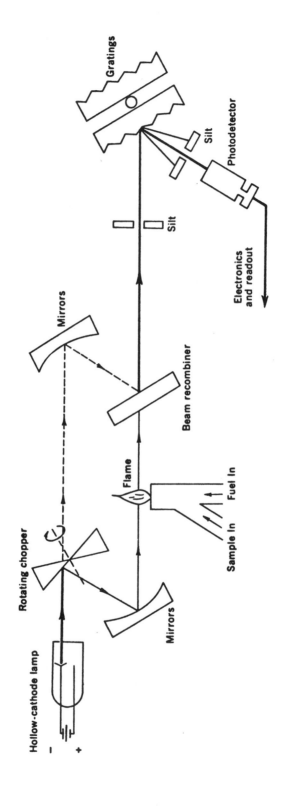

Figure 7. Perkin–Elmer model 303 atomic absorption spectrometer.

of the presence of certain substances which combine with the metal under analysis to form relatively nonvolatile compounds which do not break down in the flame. For example, calcium in the presence of phosphate exhibits this effect. This may be remedied by sequestering the calcium with EDTA. Matrix effects may be minimized by separation or by adding approximately the same amount of matrix component to the standard solution.

In contrast to flame photometry, there is very little interelement interference in atomic absorption spectrophotometry. Also, while sensitivity in flame photometry is critically dependent on flame temperature, this is not the case for atomic absorption spectrophotometry.

Over 60 elements can be determined readily by atomic absorption in the parts-per-million range without sample pretreatment and with an accuracy of 1-2%. The sensitivity can be vastly increased to the parts-per-billion range by extracting and concentrating the metal in a nonaqueous solvent and spraying it into the flame. This technique finds wide applications in the analysis of various metal species in natural and waste waters. Microgram-per-liter quantities of cobalt, copper, iron, lead, nickel, and zinc have been determined in saline waters by extraction of metal complexes with ammonium pyrolidine dithiocarbamate into methyl isobutyl ketone. An increase of about 60% in the atomizer efficiency can be achieved with the use of certain organic solvents.

In addition to its selectivity and sensitivity, atomic absorption spectrophotometry is a rapid and easy technique suitable for routine analysis, and it can be easily automated for monitoring effluent streams and waste water discharges. The detection limits of some common metals are given in Table 12.

The theory and application of atomic absorption spectrophotometry have been reported by Robinson[35] and Kahn.[22] The application of atomic absorption technique for trace analysis of inorganic ions in material and waste waters has been reported by various investigations.[6,17,43,44]

Sensitivity is highly dependent on flame characteristics. The type of gas mixture, the type of burner and gas flow rate are important factors which govern sensitivity.

Atomic Fluorescence Spectrophotometry

It appears from the above discussion that molecular fluorescence spectrophotometry offers the distinct advantages of greater sensitivity and selectivity over molecular absorption spectrophotometry. Relative to atmoic

Table 12

Atomic Absorption Detection Limits[1]

Metal	Detection Limit	Analytical Wave Length	Suggested Resolution
Silver	0.005	3281	7A
Aluminum[2]	0.1	3093	2A
Arsenic[3]	0.1	1937	7A
Boron[2]	6.0	2497	7A
Barium[2]	0.05	5536	4A
Beryllium[2]	0.002	2349	20A
Bismuth	0.05	2231	2A
Calcium	0.002	4227	13A
Cadmium[3]	0.001	2288	7A
Cobalt	0.005	2407	2A
Chromium	0.005	3579	2A
Copper	0.005	3247	7A
Iron	0.005	2483	2A
Mercury	0.5	2537	20A
Potassium	0.005	7665	13A
Lanthanum[2]	2.0	3928	4A
Lithium	0.005	6708	40A
Magnesium	0.0003	2852	20A
Manganese	0.002	2795	7A
Molybdenum	0.03	3133	2A
Sodium	0.002	5890	4A
Nickel	0.005	2320	2A
Lead	0.03	2833	7A
Antimony	0.1	2175	2A
Selenium[3]	0.1	1961	20A
Silicon[2]	0.1	2516	2A
Tin[3]	0.02	2246	7A
Tellurium	0.1	2143	7A
Titanium[2]	0.1	3643	2A
Thallium	0.025	2768	20A
Vanadium[2]	0.02	3184	7A
Tungsten[2]	3.0	4008	2A
Zinc	0.002	2138	20A
Zirconium[2]	5.0	3601	2A

[1]The detection limit is given by the metal concentration
in ppm which gives a signal twice the size of the peak-
to-peak variability of the background.
[2]Nitrous oxide flame required.
[3]Indicates use of argon-hydrogen flame.

absorption spectrophotometry, however, no increase in selectivity can be gained by using atomic fluorescence since the former is virtually specific for each element. Nevertheless, it is possible to increase the sensitivity of measurements with atomic fluorescence spectrophotometry by increasing the intensity of irradiation or by increasing the amplification until the system becomes noise limited. In this sense, atomic fluorescence spectrophotometry offers greater flexibility and sensitivity than atomic absorption spectrophotometry.[15,1]

The analytical relationships in atomic fluorescence spectrophotometry can be given simply by the following equation

$$F = k\phi I_0 \ C \tag{16}$$

where k is a constant specific for a given atomizer and instrumental set of conditions. From Equation 16, the fluorescence signal is proportional to the intensity of irradiation and to the concentration of the preatomic species in the solution. A schematic diagram of an apparatus for atomic fluorescence spectrophotometry is given in Figure 8.

The technique is inherently simple and practically any flame spectrophotometer may be adapted for this purpose without interference with its normal mode of operation. A continuous source with simple monochromator may be used. Where high sensitivities in the subnanogram range are required, it is necessary to use individual spectro discharge lamps.

The sensitivity of atomic fluorescence spectrophotometry for several elements using oxygen-hydrogen and hydrogen-air flames[15] are given in Table 13.

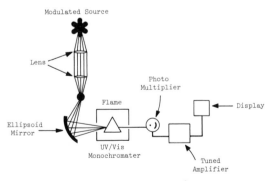

Figure 8. Apparatus for atomic fluorescence spectrophotometry.

Table 13

Limits of Detection for Several Elements in ppm

Element	H_2/O_2	H_2/Air
Ag	0.003	0.001
Co	3.50	0.18
Cu	0.10	0.018
Fe	25.00	1.8
Mg	0.18	0.004
Mn	35.00	0.04
Ni	5.00	1.0
Tl	0.30	0.07
Zn	0.35	0.01

Flame type and characteristics are extremely important
factors which significantly influence the instrument
sensitivity. These are dependent on the type of gas mixture, type of burner and gas flow rates.[21]

Infrared Absorption Spectrophotometry

This technique is essentially a molecular absorption
spectrophotometric procedure. It involves the dispersing
of a polychromatic infrared beam of light using a suitable
prism or defraction grating. Narrow wave length bands of
infrared light are isolated by a slit and allowed to pass
through the sample. The analysis is run by passing the
polychromatic light through the sample, which absorbs portions of the light. The remaining light then passes
through a prism or grating where it is dispersed into its
component wave lengths. The resultant spectrum is scanned
by a detector which records the intensity at each wave
length. The sample solution will preferentially absorb
specific wave lengths of the radiation depending on the
virbational and rotational energies of the bonds in the
molecule. Various functional groups, composed of definite
atomic configurations with definite vibrational and rotational energies, absorb strongly at characteristic wave
lengths. Accordingly, infrared absorption analysis provides a very useful tool for the characterization of organic compounds by means of functional group analysis.

The optical system of a modern double beam infrared spectrophotometer is shown in Figure 9. The optical system consists of five main sections: radiation source, sampling area, photometer, monochromator, and detector. The infrared radiation source is usually an electrically heated Nernst filament or a Globar 1000° to 1800°C. The radiation from the source is divided by mirrors M1 and M2 into two beams, a reference beam and a sample beam, which are focused into the sample area by mirrors M3 and M4.

Reference and sample beams pass through the reference cell and sampling cell, respectively. Either beam can be blocked independently by means of opaque shutters mounted on the source. The reference beam passes through the alternator and is reflected by mirrors M6 and M8 to the rotating sector mirror M7, which alternately reflects the reference beam out of the optical system and transmits the beam to mirror M9. The reference beam is intermitted at a frequency which varies from 8-15 cycles per second (depending on the type of instrument) and is focused by mirror M10 on the slit S1. Similarly, the sample beam passes through the comb and is reflected by mirror M5 to the rotating sector mirror M7, which alternately transmits the beam out of the optical system and reflects it to mirror M9, mirror M10 and slit S1.

Accordingly at any given moment, the beam focused on slit S1 is either the reference beam or the sample beam. By this arrangement the reference beam and the sample beam are combined into a single beam of alternating segments which establishes a switching frequency at the detector equal to the speed of rotation of M7. Optical null point balancing is achieved by driving the alternator in and out of the reference beam in response to the signal created at the sample beam, until both beam intensities match. It is the movement of the alternator which is recorded by the recording pen.

The combined beam passes through the monochromator entrance slit S1 to the mirror M11, which reflects it through the prism to the Littrow mirror M12. The beam is then dispersed over a range of wave lengths which are reflected back through the prism (to increase its dispersion) to mirror M11, and then to mirror M13 which focuses the beam on exit slit S2. The rotation of M12 produces a scan of a desired frequency at the exit slit S2, and, consequently, at the detector. The use of grating instead of prism for beam dispersion is becoming more widespread.

The beams coming through the exit slit S2 of the monochromator are reflected by mirror M14 to an ellipsoidal mirror M15. The foci of the ellipsoidal mirror are the

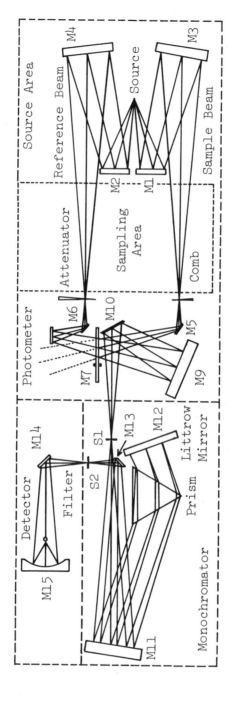

Figure 9. Perkin-Elmer Model 621 infrared spectrometer.

exit slit S2 and the detector.

Only a few solvents can be used in infrared analysis since most solvents absorb infrared very strongly. This can be seen by examining the infrared absorption chart given in Figure 10.

Interpretation of infrared spectra is a rather complicated process. There are several factors which can influence the spectra and which must be taken into account in interpretation and detection of functional groups. Substitutions and molecular symmetry will shift functional group absorption locations and will affect the intensity of the absorption bands. For example, the absorption for the triple bond in acetylenic molecules is intense for asymmetrical molecules but is weak for symmetrical ones. In addition, sample crystallinity, solute-solvent interaction, and solute-solute interaction will affect the appearance of the spectrum. Also, if the sample is prepared in solution, as a mineral oil (Nujol) mull or as a potassium bromide pellet, or if the pure sample is presented to the instrument, different spectra for the same material can result.

The infrared spectrum in the range between 0.7μ to 2.5μ provides unique analytical possibilities. This is referred to, generally, as near infrared absorption spectrophotometry. This region is primarily concerned with overtone vibrations, and thus the absorption bands are generally weaker than those in the rest of the infrared region.

The difference between near and conventional infrared instrumentation is that with near infrared an ordinary incandescent light source can be used. In addition, quartz optics (or defraction gratings) and also quartz cells for handling liquids can be used along with lead sulfide detectors. Conventional infrared covers the range between 2.5μ to 15μ and requires a source of infrared radiation, crystal optics (sodium, potassium, cesium, calcium, beryllium or silver halides, or sapphire), and bolometer or thermocouple detectors.

In spite of the fact that the analytical utility of the near infrared region is limited because of few absorption peaks, it can be readily used to determine hydroxy compounds such as alcohols, phenols and carboxylic acids as well as hydrazines, imines and similar compounds as shown in Table 14.

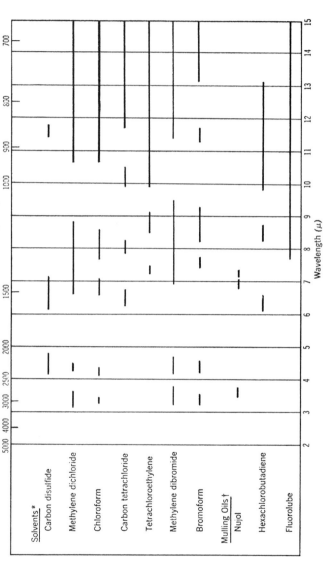

Figure 10. Transparent regions of common solvents and mulling oils (*the open regions represent solvents transmitting more than 25% of the incident light at 1mm thickness. †Open regions for mulling oils indicate transparency of thin films.)

Table 14

Spectra-Structure Correlations and Average Molar Absorptivity Data for Near-Infrared Region

Microns

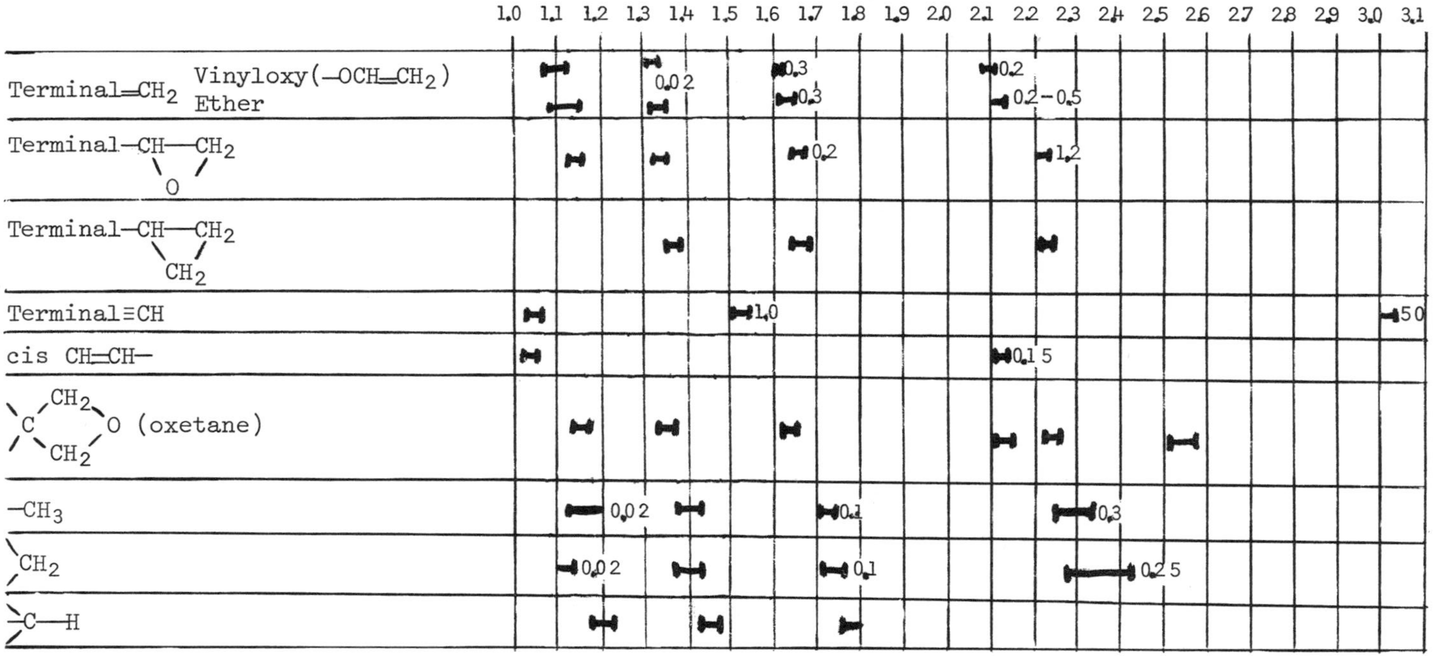

1.0 1.1 1.2 1.3 1.4 1.5 1.6 1.7 1.8 1.9 2.0 2.1 2.2 2.3 2.4 2.5 2.6 2.7 2.8 2.9 3.0 3.1
Terminal$=CH_2$ Vinyloxy($-OCH=CH_2$) Ether
0.02
0.3
0.3
0.2
0.2 - 0.5
Terminal$-CH-CH_2$ O
0.2
1.2
Terminal$-CH-CH_2$ CH_2
Terminal$\equiv$CH
1.0
50
cis CH=CH$-$
0.15
C CH_2 O CH_2 (oxetane)
$-CH_3$
0.02
0.1
0.3
CH_2
0.02
0.1
0.25
$-C-H$

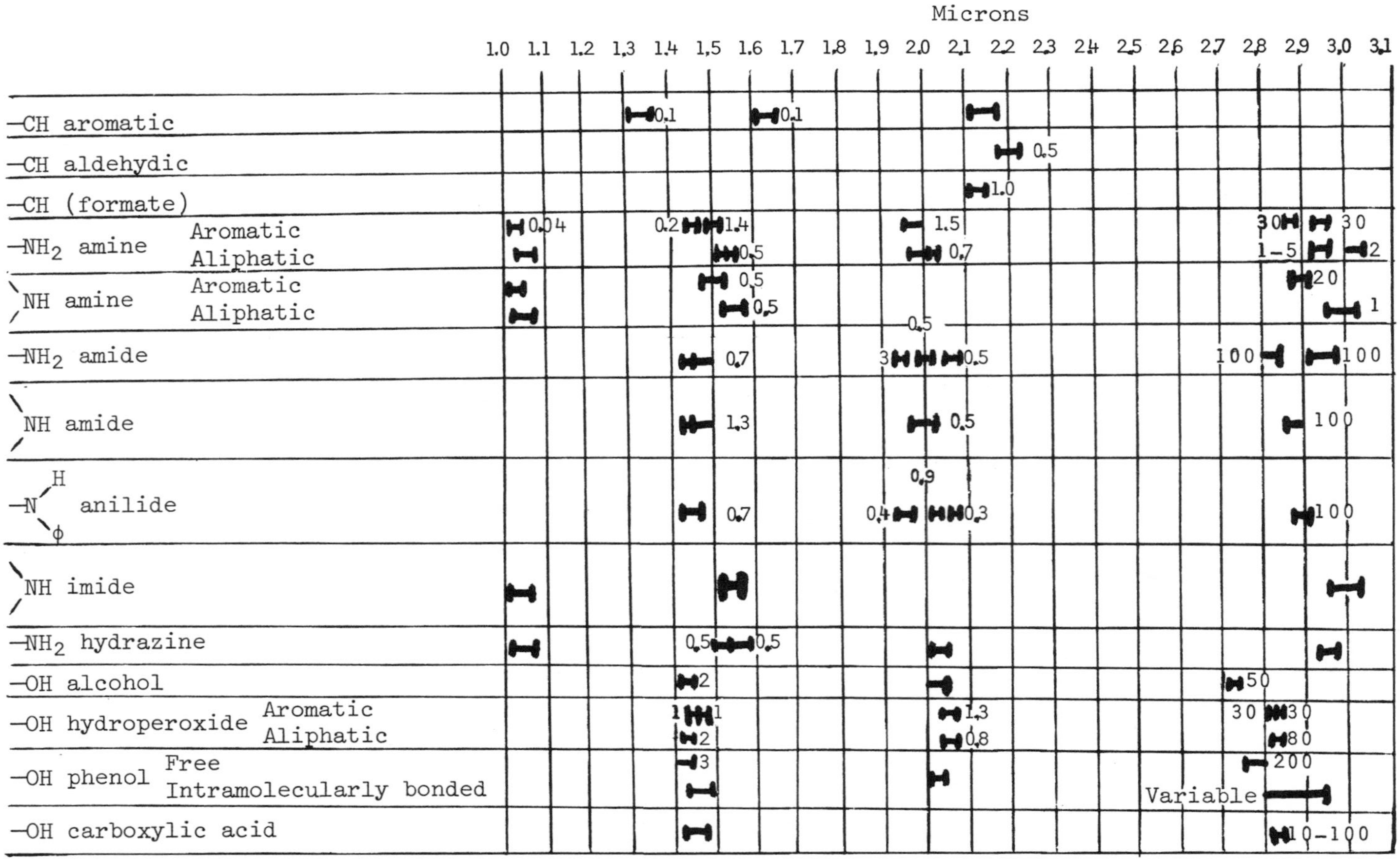

Microns

1.0 1.1 1.2 1.3 1.4 1.5 1.6 1.7 1.8 1.9 2.0 2.1 2.2 2.3 2.4 2.5 2.6 2.7 2.8 2.9 3.0 3.1

—CH aromatic

—CH aldehydic

—CH (formate)

—NH₂ amine   Aromatic / Aliphatic

NH amine   Aromatic / Aliphatic

—NH₂ amide

NH amide

N—H anilide (φ)

NH imide

—NH₂ hydrazine

—OH alcohol

—OH hydroperoxide   Aromatic / Aliphatic

—OH phenol   Free / Intramolecularly bonded

—OH carboxylic acid

Variable

Microns

*Published data, mostly obtained in CCl_4 solution. Units are liter/mole-cm.

Infrared identification of organic matter in natural and waste waters is generally preceeded by separation and concentration procedures, such as gas chromatography or liquid-liquid extraction. Liquid samples in a solvent matrix, or solid potassium bromide mulls, or paste in mineral oil may be analyzed for typical infrared absorption bonds to provide structural identification. With reasonable care, it is feasible to reproduce absorptivities within approximately 0.5% when employing the same sample cell and spectrophotometer.

A modification of infrared spectrometry, offering unique possibilities for analysis of organic matter in water, involves attenuated total reflectance.[20] This method is based on the passage of a monochromatic or monochromatically scanned light into a crystal of suitable material which is in contact with the test solution, followed by detection of the reflected light intensity. Changes in light intensity are related to changes in the type and concentration of "light active" substance in contact with the crystalline material. The principle is based on energy reflection at the interface between media of different refractive indices. Little sample preparation is required, and the use of two attenuation attachments permits differential spectral measurement.[29,48]

Ultraviolet Absorption Spectrophotometry

The principle of operation of ultraviolet absorption spectrophotometry is similar to that of the infrared technique except that a polychromatic ultraviolet light source is used. A similarity, in the case of ultraviolet spectrophotometers, is that the prism or defraction grating is placed before the sample so that only narrow wave length bands pass through the sample. A difficulty with this technique is that a large ultraviolet input of energy could cause fluorescence to occur in the test solution which could confuse the spectrum. Furthermore, large amounts of ultraviolet energy could cause photochemical reactions to occur with some materials.

Careful consideration also should be given to the selection of the solvent. Figure 11 shows UV transparency ranges for a number of solvents.

In contrast to the infrared technique, only few atomic configurations in the molecules absorb in the ultraviolet range. Those that do absorb do not have to be functional groups but rather highly resonating molecular configurations. The energy involved is due to electron transitions from one state to another. Typical examples are

88

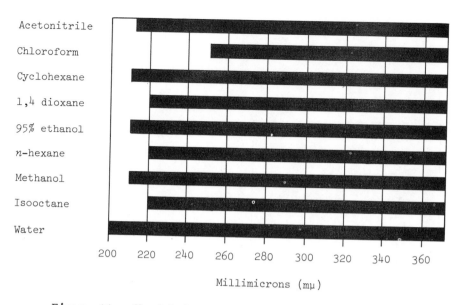

Acetonitrile									
Chloroform									
Cyclohexane									
1,4 dioxane									
95% ethanol									
n-hexane									
Methanol									
Isooctane									
Water									

200 220 240 260 280 300 320 340 360

Millimicrons (mµ)

Figure 11. Useful transparency ranges of solvents
in ultraviolet region.

conjugated saturated bonds and aromatic systems with
their conjugated cyclic structures.

Ultraviolet absorption spectra are generally much
simpler to interpret than infrared spectra because only a
few absorption peaks are obtained in the UV range. UV
spectra can be used to give some idea of the class of
compound under investigation but cannot be used to give
conclusive identification. This usage is more applicable
in cases of compounds of highly conjugated bonds such as
aromatics, where several absorption peaks exist. The
precision and accuracy of quantitative ultraviolet analysis can be obtained generally in the range of 0.2-2% at
maximum sensitivity.

A schematic diagram of the optical system of a double
beam ultraviolet spectrophotometer is shown in Figure 12.
The radiation source is a hydrogen discharge tube which
can be replaced by a tungsten incandescent lamp for
measurement in the visible region. The light emitted
from either source is focused by mirrors M1 and M2 onto
the entrance of slit S1 of the monochromater and is
then dispersed into its separate wave lengths. This is
done by collimating the light beam by mirror M3 and
reflection onto prism P1. The light beam is reflected
back through the prism by a mirrored surface on the back

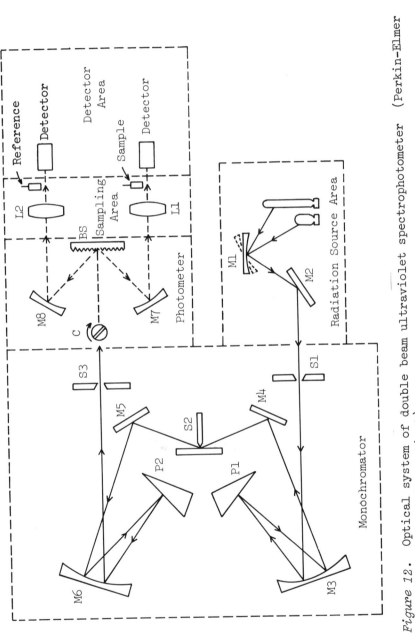

Figure 12. Optical system of double beam ultraviolet spectrophotometer (Perkin-Elmer Corp., Connecticut).

of the prism. Dispersion takes place during both passes
through the prism. The light emerging from the prism Pl
is then reflected by mirrors M3 and M4 onto slit S2.
From slit S2 the light passes through the second stage
of the monochromator (M5, M6, and P2) which is a mirror
image of the first stage.

The two prisms Pl and P2 rotate simultaneously and the
wave length of radiation at the slit S3 is controlled by
the angular positions of the prisms. The double dispersion system in this instrument offers the advantage of
minimizing stray radiation.

The monochromatic light emerging from slit S3 is pulsed
by chopper C and is split into sample and reference beams
by splitter BS. The sample and reference beams are reflected by mirrors M7 and M8 and through lenses L1 and L2
into the sample. It is important to note that optics for
the transmission of ultraviolet radiation are made of
quartz. Detection of light intensities of sample and
reference beams is done by photomultiplier tubes. The
difference in light intensities creates an off-balance
voltage which is displaced either linearly as trans-
mittance or logarithmically as absorbance.

UV absorption spectrophotometry has not been widely
used for the analysis of natural and waste waters, despite
an abundance of literature on the application of UV
visible spectrophotometry for organic analyses. Exhaustive
coverage of this subject can be found in review articles
appearing in some recent periodicals (*e.g.*, Journal of
Molecular Spectroscopy, Spectrochimica Acta, Applied
Spectroscopy, Talanta, and Analytica Chimica Acta).

Phenolic compounds in waste effluents have been determined by UV spectrophotometry using a technique based
on a comparison of the bathochromic shifts in wave lengths
in alkaline and neutral solutions.[46]   This method is
reported to be particularly useful whenever organic
separation is difficult.

UV spectrophotometry also has been used for monitoring
the composition of reaction mixtures in industrial processes and industrial waste effluents.[46] The method is
based on measuring UV absorption spectra for compounds
having aromatic or conjugated unsaturated molecular configurations. The instrument used for these analyses
utilized a mercury discharge lamp with a principal UV
radiation at 2537Å.

Electrochemical Analysis

Electrochemical methods are often well-suited for the
analysis of natural waters and waste waters. A variety

of electrode systems and electrochemical techniques have been used routinely for *in situ* analysis and continuous monitoring of waste effluents.

For purposes of this discussion, electrochemical methods are conveniently classified as being based either on the passage of faradaic current, *e.g.*, classical polarography or on electrode equilibrium, *e.g.*, potentiometry.

Methods based on passage of faradaic current

Since its development classical polarography has been used widely for water analysis. A schematic diagram of a classical polarograph is shown in Figure 13. Polarographic measurements are obtained by determining the time-averaged currents of the dropping mercury electrode (DME) under diffusion conditions. A typical polarographic current-voltage curve is shown in Figure 14. The response is described approximately by the Ilkovic equation,

$$i_d = [605 \ n \ D^{1/2} \ m^{2/3} \ t^{1/6} \ ]c \qquad (17)$$

where i_d is the average diffusion current (μ amp), t is the drop-time (sec), m is the mass rate of flow of mercury (mg/sec), D is the diffusion coefficient of the electroactive species (millimoles per liter), and n is the number of electrons per molecule involved in the electrode reaction. For a typical case in which m = 2 mg/sec,

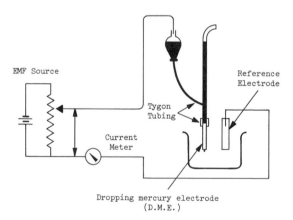

Figure 13. Schematic diagram of classical polarograph.

92

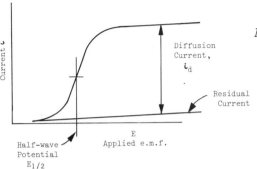

Current *i*

Diffusion
Current,
i_d

Residual
Current

E
Applied e.m.f.

Half-wave
Potential
$E_{1/2}$

Figure 14. Essential features of a polarographic current-voltage curve, alternatively known as a "polarogram" or "polarographic wave."

$t = 4$ sec, and $D = 10^{-5}$ cm^2/sec, the electrode response will be $i_d/C = 3.82$ μ amp/millimole/liter.

The potential corresponding to the midpoint of the polarographic wave, where the current is exactly one-half of the maximum diffusion current $(i_d/2)$, is known as the "half-wave potential, $E_{1/2}$". This is a characteristic property of a given redox system and is used for its identification.

As a result of the capacitance current used in charging the DME double layer, the sensitivity of classical polarography with the dropping mercury electrode is limited to approximately 10^{-5}M. However, by means of preconcentration techniques it may be possible to extend the sensitivity range significantly. Copper, bismuth, lead, cadmium, and zinc have been measured in the range of 0.01 mg/liter after extraction with dithizone and carbon tetrachloride.[21] Preconcentration by ion exchange, freeze drying, evaporation, or electrodialysis may be used.

A significant problem in the application of classical polarography to industrial waste water analysis is the interference produced by electroactive and surface active impurities. Such impurities, frequently present in waste waters, may interfere with electrode reaction processes and cause a suppression and/or a shift of the polarographic wave.[28] Modifications of polarographic techniques, such as "differential polarography" and "derivative polarography," may be used to increase the sensitivity and minimize the effect of interferences.[37]

A schematic diagram of differential polarography is shown in Figure 15. In this method the two cells, C_1 and C_2, are arranged in opposition in a bridge, and the difference of the currents through both cells is measured. R_1 and R_2 are adjustable resistances and M is the measuring instrument.

If one of the cells contains, for example, only the supporting electrolyte, and the other has in addition a depolarizer, then the measuring instrument will register only the current produced by the deposition of the depolarizer if the electrical and mechanical dimensions of both cells are identical. The advantages of this method are:

(a) *Sensitivity* is greatly increased because the capacitance current and the residual current, including that resulting from conducting salts present as impurities, are eliminated.

(b) *Separability* is also improved, especially if not only a conducting salt but also a corresponding amount of the interfering ions is added to the second cell, which contains the blank. By synchronizing exactly the two dropping-mercury electrodes, the serrations caused by the falling drops are completely eliminated from the wave, thus improving the separability.

(c) *Resolvability* is usually better because the resulting polarogram is smooth and does not have the serration caused by the dropping process.

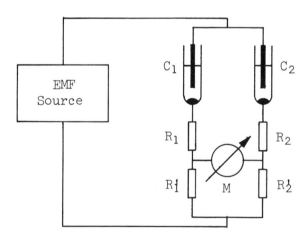

Figure 15. Schematic diagram of differential polarography.

In derivative polarography, as the name indicates, the differential quotient di/dE (or $\Delta i/\Delta E$) is plotted against E, instead of plotting i against E. This means that instead of recording the usual polarographic curve, one records its gradient. The desired relationship $\Delta i/\Delta E$ is obtained and recorded directly by a suitable circuit arrangement, or the usual i/E relationship is first measured and then electrically differentiated. The principle of the former method is shown in Figure 16. As in the case of differential polarography, two synchronous dropping-mercury electrodes are used, but both are immersed into the same solution. The ratio of the resistances R_1 and R_2 is so chosen that the two capillary electrodes have a potential difference (ΔE) of about 10 mV with respect to the opposing mercury pool. The galvanometer M measures the corresponding difference of the partial currents in R_2 and R_3 (Δi) and hence the relationship $\Delta i/\Delta E$ is obtained. In another arrangement it is possible to use a differential galvanometer.

The derivative polarogram exhibits a maximum of the half-wave potential. This gives an increase in the separability and resolvability between waves. It is important in both differential and derivative polarography that both capillaries provide completely synchronous mercury drops. This can easily be achieved by mechanically dislodging the mercury drops and by having a single mercury reservoir (same head).

Pulse polarography has the advantage of extending the sensitivity of determination to approximately 10^{-8}M. The technique is based on the application of short potential pulses of 50 m/sec on either a constant or gradually increasing background voltage. Following application of the pulse, current measurements are usually done after the spike of charging current has decayed. The limiting current in pulse polarography is larger than in classical polarography. The diffusion current equation for pulse polarography is given by the Cottrell equation,

$$i_d = [n \; F \; A \; (\frac{D}{\pi t})^{0.5}] \; C \qquad (18)$$

Derivative pulse polarography, which is based on superimposing the voltage pulse upon a slowly changing potential (about 1mv/sec) and recording the difference in current between successive drops versus the potential, is even more sensitive than pulse polarography.

Electroanalytical methods lend themselves to various modifications through variations of the current, potential and time relationships. In general, two methods have been

95

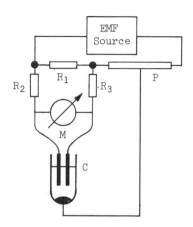

Figure 16. Deriva-
tive polarography
schematic diagram.

found most practical, chronoamperometry and chronopotentiometry. The former technique is based on controlling
the potential and measuring the variations of electrolysis
current with time. This is done at static or dynamic potentials which vary as a function of time. In the latter
technique the electrolysis current and the potential are
measured as a function of time. The current may be held
constant or allowed to vary in a known fashion with time.

A generalized classification of electroanalytical procedures, reported by Charlot,[10] is shown in Table 15. The
table includes amperometric, potentiometric and coulometric
titration techniques. For example, in potentiometric
titrations $E = f(v)$, and in a coulometric titration
$E = f(Q)$, where v and Q are the volume of titrant and
amount of electricity, respectively.

Methods which make use of a cathode ray oscilloscope
are usually called oscillographic polarography. Measurements are usually either chronoamperometric or chronopo-
tentiometric in nature. Cathode-ray polarography or
oscillographic polarography has been used for analysis of
natural waters and waste waters, with a reported sensitivity of $10^{-7}M$.[27] This technique uses a cathode-ray
oscilloscope to measure the current-potential curves of
applied (saw-tooth) potential with rapid sweeps during the
lifetime of a single mercury drop. Multiple sweep techniques are also applicable. The peak current (i_p) in the
resulting polarogram is related to the concentration of
the electroactive species for a reversible reaction in
accordance with the Randles-Sevcik equation,

$$i_p = [k \, n^{3/2} \, m^{2/3} \, t^{2/3} \, D^{1/2} \, v^{1/2}]C \qquad (19)$$
$$\text{(reversible process)}$$

or

Table 15

Classifications of Electrochemical Procedures

Imposed	Measured	Name
c = const.	$i = f(E)$ or $E = f(i)$	voltammetry
i = const. (o or l_ε)	$E = f(c)$ or $\Delta E = f(c)$	potentiometric titrations (with one or two indicator electrodes)
E = const. or ΔE = const.	$i = f(c)$	amperometric titrations (with one or two indicator electrodes)
c = const. $E + \Delta E\sim$	$i \sim i.e.$ $\frac{di}{dE} = f(E)$	AC polarography sine-wave polarography
c = const. $E + \Delta E_\sqcap$	$i_\sqcap i.e.$ $\frac{di}{dE} = f(E)$	square-wave polarography
$\Delta E\sim$ or $\Delta E_\sqcap$	$\frac{di}{dE} = f(c)$ ($i = o$)	AC amperometric titrations

Steady state $f(i,E,c) = 0$

Steady state + periodic state of small amplitude

Steady state + periodic state (*continued*)

$\Delta i\~$ or $\Delta i\~_{\mathfrak{l}}$	$\dfrac{dE}{di} = f(c)$ ($i = o$)	AC potentiometric titrations
c = const. E = const.	$i = f(t)$	chronoamperometry
c = const. $E = E_i + vt$	$i = f(t)$ or $i = f(E)$	linear chronoamperometry
c = const. i = const.	$E = f(t)$	chronopotentiometry
i = const.	$E = f(t)$ $\tau = f(c)$	chronopotentiometric titrations
c = const. E = const. + vt (saw tooth)	$i = f(t)$ or $i = f(E)$	oscillographic polarography: with saw-tooth voltage
c = const. $E = E\~$	$i = f(t)$ or $i = f(E)$	with sine-wave voltage
c = const. $i = i\~$	$E = f(t)$ or $\dfrac{dE}{dt} = f(t)$ or $\dfrac{dE}{dt} = f(E)$	with sine-wave current
c = const. $i = i_0 + \Delta i\~$	$E\~i.e.$ $\dfrac{dE}{dt} = f(t)$	AC chronopotentiometry

not periodic

periodic (large amplitude)

Nonsteady state $f(t,E,c) = o$

Nonsteady state + periodic state of small amplitude

$$i_p = [k \ m^{2/3} \ (\alpha_1 n_a)^{1/2} \ t^{2/3} \ D^{1/2} \ v^{1/2}]C \qquad (20)$$
$$(\text{irreversible process})$$

where $v = \dfrac{dE}{dt}$ is the voltage sweep, α is the transfer co-efficient and n_a is the number of electrons transferred in the rate determining reaction.

Oscillographic polarography has the advantages of relatively high sensitivity, high resolution, and rapid analysis. Traces of Cu, Pb, Zn and Mn can be determined at 0.05 mg/ml levels in natural waters by this technique.[27] A typical current voltage curve is shown in Figure 17. The peak potential E_p is related to the conventional polarographic half-wave potential by the following equation

$$E_p = E_{1/2} - 1.1 \ \frac{RT}{nf} \qquad (21)$$

Accordingly E_p appears at 28/n mv more negatively than $E_{1/2}$.

One of the most interesting electrochemical approaches to trace metal analysis in aquatic environments is anodic stripping voltammetry.[40] This technique involves two consecutive steps: (a) the electrolytic separation and concentration of the electroactive species to form a deposit or an amalgam on the indicator electrode, and (b) the dissolution (stripping) of the deposit. The separation step, best known as the pre-electrolysis step, may be performed quantitatively or arranged to separate a reproducible fraction of the electroactive species. This can be done by performing the pre-electrolysis step under carefully controlled conditions of potential, time of electrolysis, and hydrodynamics of the solution. The deposition under reproducible stirring conditions follows the first order kinetic expression

$$Q_t = Q_0(1 - e^{-kt})\ldots\ldots\ldots \qquad (22)$$

where Q_t is the number of coulombs in the stripping peak at any time, Q_0 is the number of coulombs equivalent to the total metal ion in the sample solution, t is the deposition time (sec), and k is the rate constant. The rate constant, k, is equal to

$$k = \frac{DA}{V\delta}\ldots\ldots\ldots\ldots \qquad (23)$$

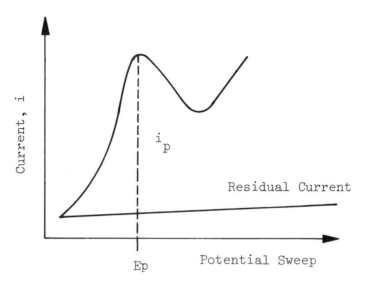

Figure 17. Essential features of a cathodic chronoamperometric curve.

where D is the diffusion constant of electroactive species
in the test solution in cm^2/sec, A is the electrode surface area in cm^2, V is the volume of the solution in cm^3,
and δ is the effective thickness of diffusion layer in
cm.

Ordinarily partial deposition of a sample is sufficient
for analysis. Under conditions where the rate constant
is not reproducible for different samples, extension of
the plating increases the accuracy of the determination.
With a long deposition time, the expression e^{-kt} approaches
zero; therefore, slight changes in k which may occur between analyses under adverse working conditions, *e.g.*,
onboard research vessels, have little effect on the precision of the results. The deposition step can be
arranged to segregate the species present as well as
deposit a reproducible fraction of the sample.

Quantitative determination of the deposited materials
is made in the stripping step, most commonly by a chronoamperometric technique with a linear potential sweep.
Commercial polarographic equipment employing operational
amplifiers may be used for slow potential sweep rates, in
the order of 2 to 100 mv/sec. When the proper electrode
is used under controlled conditions, the concentration of
metals in solution is linearly proportional to the peak
current, and qualitative identification of the metal ions

in solution is obtained from peak potential values.

The stripping step is usually done in an unstirred solution by applying a potential--either constant or varying linearly with time--of a magnitude sufficient to drive the reverse electrolysis reactions. Quantitative determinations are done by integrating the current-time curves (coulometry at controlled potential) or by evaluating the peak current (chronoamperometry with potential sweep). Several modifications of the separation and stripping steps have been reported.[39]

Hanging-drop mercury electrodes of the Gerischer's[19] or Kemula's type[23] have been widely used for anodic (or cathodic) stripping analysis. Greater sensitivity has been achieved by use of electrodes which consist of a thin film of mercury on a substrate of either platinum, silver, nickel, or carbon.[30] Errors due to nonfaradaic capacitance current components can be minimized by proper choice of stripping technique.

The main advantage of stripping voltammetry is its applicability to trace analysis. The technique has been applied for metal analyses in sea water,[4] natural waters and waste waters.[27,2]

Measurement of Current-Potential Curves

The underlying principle in all electroanalytical methods based on passage of faradaic current is the current-potential relationship. Current and potential relationships are measured at either steady state conditions or transient conditions where they vary with time. Generally speaking, there are two basic types of electroanalytical systems used in this connection. The first utilizes a two-electrode cell while the second utilizes a three-electrode assembly.

The two-electrode system is shown in Figure 18. It consists of an indicator electrode and a reference electrode. The circuit consists of an electrolysis cell, an adjustable voltage source, an electronic voltammeter to measure the potential difference between the electrodes, and a microammeter to measure the electrolysis current.

The measured potential difference between the indicator and reference electrodes includes the ohmic drop due to resistance of the cell, $i.e.$,

$$V = (E_A - E_C) + |i| \ R \qquad (24)$$

where V is the measured potential, E_A and E_C are the potentials of the anode and cathode, respectively, $|i|$ is the

101

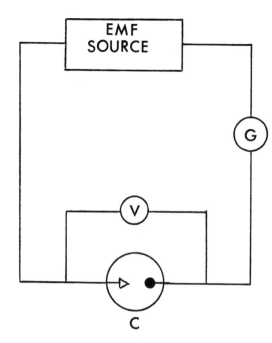

Figure 18. Basic electrical circuit for two electrode system. C = electrolysis cell, G = microammeter, V = millivoltmeter, ●—— = indicator electrode, ——▷ = reference electrode.

current passing in the cell and R is the cell resistance. If R and i are both small, *e.g.*, R < 1000 ohms and i<10 µa, the ohmic drop is less than a millivolt and can be neglected. In certain applications, however, the resistance of the cell is several thousand ohms and should be corrected for.

The three-electrode system shown in Figure 19 consists of an indicator electrode, a reference electrode and an auxilliary electrode (sometimes referred to as the counter electrode). Instead of measuring the potential difference between the two electrodes used to produce the electrolysis, one measures the potential difference between the indicator electrode and the reference electrode. Under these conditions the electrolysis current passes between the indicator and auxilliary electrodes. Since virtually no current passes between the indicator and the reference electrode, there should be no ohmic drop, and the measured potential is the exact potential of the indicator electrode.

The three electrode cell configuration lends itself to a variety of controlled potential or controlled current electroanalytical techniques. Schematic diagram of circuit arrangements is shown in Figure 20. Operational amplifier circuits have been advised to provide controlled potential and controlled current voltammetric techniques. Figure 21 shows a schematic diagram of an operational amplifier controlled potential system. Amplifier A serves

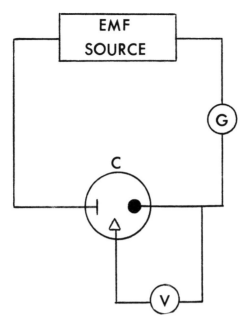

Figure 19. Circuit arrangement for controlled current and controlled potential voltammetry.

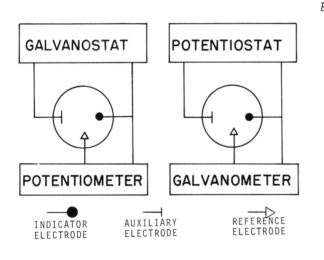

INDICATOR ELECTRODE   AUXILIARY ELECTRODE   REFERENCE ELECTRODE

Figure 20. Basic electrical circuit for three electrode system. C = electrolysis cell, G = microammeter, V = millivoltmeter, ●━ = indicator electrode, ━▷ = reference electrode, and ━┥ = auxilliary electrode.

103

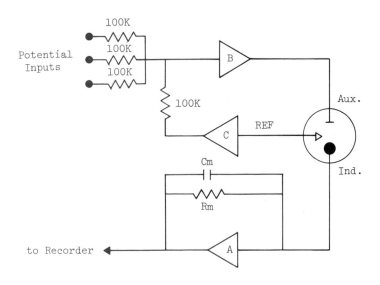

Figure 21. Operational-amplifier controlled potential
            system.

to maintain the potential of the indicator electrode at
ground (input potential for amplifier), less the IR drop
in the electrode and its connecting leads. It also serves
to provide a signal proportional to the current flowing
between the indicator electrode and auxilliary electrodes
$(e_o = IR_m)$. Amplifier B is a follower amplifier which
responds to the potential of its input relative to ground
(the input of Amplifier A). Accordingly, the output of
the follower amplifier will be equal to the potential of
the reference electrode relative to the indicator electrode, less the IR drop between these points. The stable
state for the loop comprising Amplifier C and the follower
corresponds to ground potential at the input of Amplifier
C. In case all resistors connected to this input are
identical, the stable state is achieved when the output of
the follower amplifier is equal and opposite in polarity
to the sum of the potentials applied at the input resistors.
Consequently, the potential of the indicator electrode
relative to the reference electrode will be equal to the
sum of the potential inputs (same polarity) less the IR
drop between the input of Amplifier A and the input of the
follower amplifier. Because essentially no current flows
through the reference electrode, the IR drop within the
reference electrode compartment is negligible, even with
low-conducting solutions.

A variety of excellent discussions on the principle of operational amplifiers and their application in electro-analytical chemistry can be found in the literature. Of particular interest are articles by Reilley,[34] Booman and Holbrook,[7] Schwarz and Shain,[38] and Smith.[42]

A variety of "multipurpose electroanalytical" instruments are now available. These are versatile instruments applicable to many distinct electroanalytical procedures. Its modular construction permits its use for voltage-scan voltammetry, current-scan voltammetry, chronoamperometry, chronopotentiometry, coulometric analysis, controlled potential electrolysis, and potentiometric, amperometric and conductometric titrimetry.

The first of these multipurpose instruments was described by D. D. Deford at the 133rd National American Chemical Society, San Francisco, 1958.[13] Since then, various innovations and modifications were introduced, primarily by Shain,[40] Lauer, Schlein, and Osteryoung,[25] Morrison[31] and Ewing and Brayden.[16]

Methods Based on Electrodes at Equilibrium

Electroanalytical methods based on electrode equilibrium include a variety of membrane electrode systems which are applicable for measurement of hydrogen ions, metal ions and inorganic anions in natural waters and waste waters. A typical example is that of the glass electrode for pH measurement.

Recent developments in glass electrodes make it possible to use these electrodes in analysis for certain metal ions, particularly sodium and potassium.[14] The doping of ordinary glass pH electrodes with Al_2O_3 greatly enhances the "alkaline error" or the response of these electrodes to alkalies, although at the same time reduces their pH response.

It has been found experimentally[14] that for a mixture of two cations, *e.g.*, Na^+ and K^+, the behavior of a modified glass electrode may be described by a modified form of the Nernst equation

$$E = \text{constant} + \frac{2.3\ RT}{nF} \log\left[a + \Sigma\ K_i a_i^{n/n_i}\right] \qquad (25)$$

where E is the electrode potential, 2.3 RT/F is the Nernst factor, n is the charge of the ion being measured, a is the activity of the ion being measured, K_i is the selectivity coefficient for the ith interfering ion, a_i is the activity and n_i is the charge of the ith interfering ion. By varying the composition of the glass in the system $M_2O\cdot Al_2O_3\cdot SiO_2$—where M_2O is either Li_2O, Na_2O, K_2O,

Rb_2O or Cs_2O--it is possible to vary the selectivity of response of the glass to each of the various alkali metal ions.

Liquid ion exchange has been used in conjunction with potentiometric specific-ion electrode systems. One example is the calcium-selective membrane electrode,[36] which consists of the calcium salt of dodecylphosphoric acid dissolved in di-n-octylphenyl phosphate. This liquid ion-exchanger is immobilized in a porous inert membrane such as cellulose. Below a solution concentration of 10^{-6} moles Ca^{+2} per liter, the membrane potential is a constant, independent of the calcium ion activity. This has been attributed to the organic calcium salt solubility, which maintains a constant limiting activity of calcium ions at the membrane solution interface.[36,3] Otherwise, the behavior of the liquid ion exchange membrane electrode is similar to that of the glass electrode, which may be expressed in terms of the modified Nernst relationship given in Equation 25.

Single crystal membrane electrode systems (solid-state membrane electrodes) recently have been applied to water analysis.[3,18] The fluoride electrode[18] which is made of lanthanum fluoride crystal membrane doped with a rare earth which presumably acts in such a way that only fluoride carries the current across the membrane, is a good example. This electrode system is highly selective for fluorides, but at high pH hydroxide ions constitute a major interference which limits its usefulness in this range. The introduction of anion selective, precipitate-impregnated membrane electrodes[32,33] has been an important recent development for electrochemical analytical techniques. The membrane is made of a silicone rubber matrix which incorporates precipitated particles of silver halide, or barium sulfate. The electrode is relatively insensitive to the nature of the cations. The most selective and sensitive electrode system of this type is the silver iodide membrane electrode which responds to iodide concentrations as low as $10^{-7}M$, with relatively little interference from common ions.[33]

From Equation 25 it is apparent that potentiometric membrane electrode systems are sensitive to the activity of the electroactive species. In order to use such electrode systems for the determination of concentrations rather than activities, it is important to consider the effects of ionic strength on the activity coefficient of the electroactive species and the liquid junction potential between the test solution and the reference electrode. To avoid the uncertainty of estimating an activity coefficient, it is useful to determine the effect of an added known amount of the test species on the potential, or to adjust the ionic strength of the sample to that of a standard solution. Since the total ionic strength of the solution determines the activity coefficient for a specific ion, the activity coefficient of the ion being analyzed in the test sample will be identical to that in the standard solution. A constant ionic strength can be obtained by using a "swamping electrolyte." This technique, frequently referred to as the "ionic medium" method, has been effectively used to calibrate potentiometric membrane electrode systems for the analysis of natural waters and waste waters.[3,41]

Methods in Which Faradaic Current is Unimportant

A typical example of this type of measurement is electrolytic conductance which is commonly used to give a gross approximation of the ionic strength and salinity of natural and waste waters. The electrolytic conductance of a solution depends on the number, size, and charge of the ions present and also on some of the characteristics of the solvent, such as viscosity. The conductance L, of a solution can be represented by the expression

$$L = K_c \sum_{i}^{n} C_i \, \lambda_i \, Z_i \qquad (26)$$

where K_c is a constant characteristic of the geometry and size of the conductance cell, C is the molar concentration of the individual ions in the solution, λ_i is the equivalent ionic conductance of individual ions, and Z is the ionic charge of the individual ions. From Equation 26, it is implied that changes in the conductance of a given solution may result from chemical transformations that produce different ionic species of different charge, size or number. In addition, two solutions of the same electrical conductance do not have to be of the same chemical composition. Accordingly, electrical conductivity measurements of natural and waste waters should be critically evaluated before any assumption can be made on the chemical composition.

In oceanographic measurements, electrical conductivity is roughly proportionate to salinity. Salinity is defined in terms of chlorinity (halide concentration in parts per mille) by the expression

$$S\,^o/oo = 1.805 \; Cl\,^o/oo + 0.030 \qquad (27)$$

This correlation is usually appropriate because over 97% of the sea water in the world has a salinity between 33%

and 37%. In estuarine waters and fresh and polluted aquatic environments this correlation is less reliable.

In situ measurements of electrical conductance can be done either by "contact electrodes" or by "noncontact electrodes." In the former case, which is more common, two or more electrodes are immersed in the solution and an alternating current is passed between them. The main problem with this type of measurement is the possible polarization of the electrodes and the effect of electroactive or surface active impurities which may cause electrode "poisoning." The noncontact electrode method, also known as "inductive" or "high frequency admittance" procedure, is based on isolating the electrodes by a layer of glass or other insulating material from the test solution. A very high frequency current, usually in the megacycle per second range, is necessary. This technique eliminates electrode "poisoning" but polarization, in spite of being reduced significantly by high frequency current, is still present. In a new innovation, this difficulty is eliminated by making the test solution form a conducting loop in a transformer[12] as shown in Figure 22. The two transformers T_1 and T_2 have cores of a high-permeability metal. The oscillator O excites an alternating flux in T_1, which induces current in the loop of sea water within the glass tube. This current similarly excites T_2, which initiates a signal amplified by A and detected by a readout D. The secondary winding on T_1, with x turns, produces a current which passes in resistance RB and a winding of y turns on T_2. The direction of the current through T_2 is chosen to oppose the flux generated by the sea water loop. Upon adjusting RB so that no signal is detected by D, the resistance of the sea water loop equals RB/xy. Cox[12] also described an *in situ* inductive cell also based on the use of a sea water loop in a transformer.

Measurement of electrical conductance is usually based on the use of conductivity bridges (Wheatstone Bridge). The bridge circuit is always arranged so that at the balance point the detector indicates either zero or minimum potential. Cox[12] described a number of modifications and instrument arrangements for salinity measurements in sea waters. Blaedel and Malmstadt[5] discussed the application of the high frequency inductive techniques for conductometric titrations.

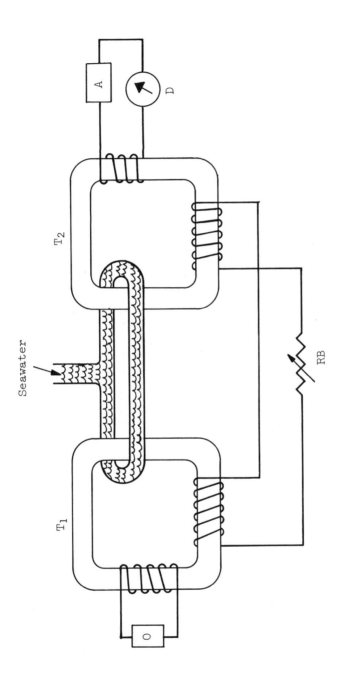

Figure 22. "Inductive-type" conductivity cell.

REFERENCES

1. Alkemade, C.T.J. *Xth Colloquium Spectroscopicum Internationale Proceedings*, E. R. Lippincott and M. Margoshes, Eds. (Washington, D.C.: Spartan Books, 1963).
2. Allen, H.E., W.R. Matson and K.H. Mancy. J. Water Pollu Control Federation 42, 573 (1970).
3. Andelman, J.B. *Proc. 140th Annual Conference, Water Pollution Control Federation*, Washington, D.C. (1967).
4. Ariel, M., and U. Eisner. J. Electroanal. Chem. 5, 362 (1963).
5. Blaedel, W.J., and H.V. Malmstadt. Anal Chem. 23, 1413 (1950).
6. Boettner, E.A., and F.J. Grunder. in *Advances in Chemistry Series*, Vol. 73, (Washington, D.C.: American Chemical Society, 1967).
7. Booman, G.L., and W.B. Holbrook. Anal. Chem. 35, 1793 (1963).
8. Bowen, E.J., and F. Wokes. *Fluorescence of Solutions* (Longmans, Green and Co., 1953).
9. Bramer, H.C., M.J. Walsh and S.C. Caruso. Abstracts of 151st American Chemical Society Meeting, Pittsburgh (March 1966).
10. Charlot, G., J. Badoz-Lambling and B. Tremillion. *Electrochemical Reactions* (Amsterdam: Elsevier Publishing Company, 1962), p 367.
11. Christman, R.F., and M. Ghassemi. J. Am. Water Works Assoc. 58, 723 (1966).
12. Cox, R.A. "Physical Properties of Sea Water" in *Chemical Oceanography*, J.P. Riley and G. Skirrow, Eds. (New York: Academic Press, 1965), p 73.
13. Deford, D.D. Abstracts 133rd ACS Meeting, 16B, San Francisco (1958).
14. Eisenman, G. in Reilley's *Advances in Analytical Chemistry and Instrumentation*, Vol. 4, (New York: Interscience Publishers, 1965), p 213.
15. Ellis, D.W., and D.R. Demers. in *Trace Inorganics In Water, Advances in Chemistry Series*, Vol. 73, (Washington, D.C.: American Chemical Society, 1967).
16. Ewing, G.W., and T.H. Brayden. Anal. Chem. 35, 1826 (1963).
17. Fishman, M.J., and M.R. Midgett. in *Trace Inorganics In Water, Advances In Chemistry Series*, Vol. 73, (Washington, D.C.: American Chemical Society, 1967).
18. Frant, M.S., and J.W. Ross. Science 154, 1553 (1966).
19. Gerischer, H. Z. Physik. Chem. (Leipzig) 202, 302 (1953).

20. Harrick, N.J. J. Phys. Chem. 64, 1110 (1960).
21. Heller, K., G. Kuhla and F. Machek. Mikrochemie 18, 193 (1935).
22. Kahn, H.L. in *Trace Inorganics in Water, Advances in Chemistry Series*, Vol. 73, (Washington, D.C.: American Chemical Society, 1967).
23. Kemula, W., A. Kublick and Z. Galus. Nature 184, 1795 (1959).
24. Laitinen, H.A. "Trace Characterization by Electrochemical Methods," in *Trace Characterization - Chemical and Physical, NBS Monograph 100* (1967), p 75.
25. Laurer, G., H. Schlein and R.A. Osteryoung. Anal. Chem. 35, 1789 (1963).
26. Macchi, G. J. Electroanal. Chem. 9, 920 (1965).
27. Maienthal, E.J., and J.K. Taylor. in Gould's *Trace Inorganics in Water, Advances in Chemistry Series*, **Vol.** 73, (Washington, D.C.: American Chemical Society, 1968), p 172.
28. Mancy, K.H., and D.A. Okun. Anal. Chem. 32, 108 (1960).
29. Mark, H.B. Chemistry Department, The University of Michigan, Ann Arbor, Michigan, private communications (1970).
30. Matson, W.R., D.K. Roe and D.E. Carritt. Anal. Chem. 37, 1954 (1965).
31. Morrison, C.F., Jr. Anal. Chem. 35, 1820 (1963).
32. Pungor, E., K. Toth and J. Havas. Acta Chim. Acad. Sci. Hung. 41, 239 (1964).
33. Rechnitz, G.A., M.R. Kiesz and S.B. Zamochnick. Anal. Chem. 38, 973 (1966).
34. Reilley, C.N. J. Chem. Ed. 11, A853 (1962).
35. Robinson, J.W. (New York: Marcel Dekker, Inc., 1966).
36. Ross, J.W. Science 156, 1378 (1967).
37. Schmidt, H., and M. von Stackelbert. *Modern Polarographic Methods* (New York: Academic Press, 1963).
38. Schwarz, W.M. and I. Shain. Anal. Chem. 35, 1770 (1963).
39. Shain, I. "Stripping Analysis," in Kolthoff and Elving's *Treatise on Analytical Chemistry*, Part I, Vol. 4 (New York: Interscience Publishers, 1959), p 50.
40. Shain, I., and S.P. Perone. Anal. Chem. 33, 325 (1961).
41. Sillen, L.G. "Master Variable and Activity Scales," in Gould's *Equilibrium Concepts in Natural Water Systems, Advances in Chemistry Series*, Vol. 67, (Washington, D.C.: American Chemical Society, 1967).

42. Smith, D.E. Anal. Chem. 35, 181 (1963).
43. Thompson, M.E., and J.W. Ross. Science 154, 1643 (1966).
44. Unterkolfer, W.L., and I. Shain. Anal. Chem. 35, 1778 (1963).
45. Weber, G., and F.W.J. Teale. trans. Faraday Soc. 53, 646 (London, England: 1957).
46. Wexler, A.S. Anal. Chem. 35, 1936 (1963).
47. White, C.E., and A. Wiessler. Anal. Chem. 34, 81R (1962), 36, 116R (1964), 38, 155R (1964), 38, 155R (1966).
48. Wilks Scientific Corp. Publication, South Norwalk, Conn. (1967).

SELECTED SUPPLEMENTARY REFERENCES

Molecular Absorption Spectrophotometry

1. Bauman, R. P. *Absorption Spectroscopy* (New York: John Wiley and Sons, 1962).
2. Charlot, G. *Colorimetric Determination of Elements* (Amsterdam: Elsevier Publishing Company, 1964).
3. Siggia, S., and H. J. Stolten. *An Introduction to Modern Organic Analysis* (New York: Interscience Publishers, 1956).

Molecular Fluorescence Spectrophotometry

1. Radley, J. A., and J. Grant. *Fluorescent Analysis in Ultraviolet Light*, 4th ed., (Princeton, N.J.: D. van Norstrand Co., Inc., 1954).
2. Strobel, H. A. *Chemical Instrumentation* (Reading, Mass.: Addison-Wesley Publishing Co., Inc., 1960).

Atomic Fluorescence Spectrophotometry

1. Alkemade, C. T. *Xth Colloquim Spectroscopicum Internationale Proceedings*, E. R. Lippincott and M. Margoshes, Eds. (Washington, D.C.: Spartan Books, 1963).
2. Winefordner, J. D., M. L. Parsons, J. M. Mansfield, and W. J. McCarthy. Anal. Chem. 39, 436 (1967).
3. *ibid*, Spectrochim. Acta 23B, 37 (1967).

Atomic Absorption Spectrophotometry

1. Elwell, W. T., and A. F. Gidley. *Atomic Absorption Spectrophotometry* (New York: The Macmillan Co., 1962).
2. Robinson, J. W. *Atomic Absorption Spectroscopy* (New York: Marcel Dekker, Inc., 1966).

Infrared Spectrophotometry

1. Bellamy. *The Infrared Spectra of Complex Molecules* (New York: John Wiley and Sons, 1958).
2. Mellon. *Analytical Absorption Spectroscopy* (New York: John Wiley and Sons, 1950).
3. Szymanski, H. A. *Infrared Handbook* (New York: Plenum, 1962).

Ultraviolet Spectrophotometry

1. Rao, C. N. R. *Ultraviolet and Visible Spectroscopy* (Butterworth and Co., Ltd, 1961).
2. West, W. *Techniques of Organic Chemistry*, Physical Methods by A. Weissberger, ed., Vol I, part III (New York: Interscience Publishers, 1960).

Electrochemical Analysis

1. Milazzo, G. *Electrochemistry* (Amsterdam: Elsevier Publishing Company, 1963).
2. Charlot, G., J. Badoz-Lambling, and B. Tremillon. *Electrochemical Reactions* (Amsterdam: Elsevier Publishing Company, 1962).
3. Reilley, C. N., and R. W. Murray. *Electroanalytical Principles* (New York: Interscience Publishers, 1963).

Chapter VI

Continuous Monitoring Systems

The development of equipment capable of continuous monitoring of the environment greatly extends man's capability for understanding and controlling water resources. The number of grab samples which can be collected and analyzed is small. Monitoring equipment provides analyses continuously or at short time intervals on a 24-hour basis. Systems for continuous monitoring are generally located at a fixed position in either a permanent building or a trailer by a stream. Therefore, the excellent coverage of fluctuations in water quality by these continuous monitoring systems will often need to be coupled with sampling programs to define the types of wastes being introduced and their dispersion characteristics. Two types of modular package continuous monitoring systems will be discussed: (1) systems which make measurements without chemical modification of the sample, and (2) systems which make measurements after suitable chemical reactions have taken place. Data collection and sensor systems will be discussed.

Systems Not Requiring Reagent Additions

A number of systems are commercially available which are designed to automatically measure and record water quality parameters by instrumental methods of analysis without the addition of reagents. These measurements include pH, conductivity, temperature, dissolved oxygen (D.O.), turbidity, oxidation-reduction potential (ORP), chloride, and solar radiation. Performance characteristics for these sensors are listed in Table 16.[9]

Table 16

Performance Characteristics of Sensors
for Continuous Monitor Systems*

Parameter	Range		Response		Accuracy		Depth	
	From	To	From	To	From	To	From	To
Temperature	-10°C-+50°C	0°C-100°C	40% in 10 sec	2 min	±0.01°C	±0.1°C	submerged	10,000 ft
D.O.	0.5 - 15 ppm	0 - 50 ppm	90% in 6 sec	90% in 20 sec	±1% of scale	±5% of scale	submerged	500 ft
Conductivity	0 - 1400 millimhos/cm	2.5 mhos/cm	–	–	±0.05% of scale	±1% of scale	submerged	1,000 ft
pH	0	14	98% in 2 sec	98% in 20 sec	±0.005 pH	±0.1 pH	submerged	200 ft
Solar Radiation	0 - 2.8 cal/cm²/min	0-5 cal/cm²/min	98% in 20 sec		±1% of scale		0 - 20 ft	–
Oxidation Reduction Potential	0 - ±1400 mv	–	–	–	±1% of scale		–	–
Chloride	1 - 10,000 PPM		90% in 20 sec		–	–	–	–
Turbidity	0 - 1,000 JCU	0 - 10,000 JCU	30 sec		±5% of scale	±2% of scale	submerged	200 ft

*From Reference 9.

Although these instruments are portable, they are generally used in a fixed location either within a plant or on the banks of a stream or they are placed in a trailer to permit convenient movement of the instrument from one location to another and to provide shelter for the instrument. Such a mobile facility will require a generator to provide the necessary electrical power or its use will be limited to locations where electrical transmission lines are nearby. Once installed, the instrument is often capable of virtually unattended operation for a period of several weeks.

The monitoring instruments are modular in design and contain a sampling module which contains the sensors, a signal conditioning module, and a data logging or transmission module.

A continuous flow of sample is provided to the sensors by a submersible pump. By using a submersible pump with a minimum flow of 10 gallons per minute, water is pushed through the piping and into the sample tank. Care must be taken to assure that no air pockets are present or dissolved oxygen and other measurements will be unreliable. The sample intake should be shielded with a screen to prevent debris from clogging the system or damaging the pump. Flexible hose should be used to connect the pump and sample tank for temporary installations while rigid pipe may be used at permanent sites.

The sampling and sensor module holds the conductivity, D.O., temperature, ORP, pH, and chloride sensors. Water is distributed to individual measuring chambers, which may be made of copper to inhibit algal growths. A high flow is presented across the face of the dissolved oxygen electrode to provide optimum response, while in the other chambers baffles prevent build-up of particulate matter by agitation of the heavy particles into the overflow drain.

Each sensor in the instrument is supplied with its own signal conditioner which converts the input to a standard electrical output. Calibration of the sensor signals is accomplished by adjustment of the controls on the signal conditioner. The signals can then be recorded on strip chart recorders or can be telemetered, stored and processed by methods described in the Data Collection section of this chapter.

Systems Requiring Addition of Reagents

The Technicon AutoAnalyzer* described in Chapter VII can be used for continuous analysis by elimination of the

---

*Registered trademark of Technicon Instruments, Inc., Tarrytown, N.Y.

sampler module. The sample may be provided by placing the line for the sample pump tube directly into the stream to be sampled. If this requires an excessive length of tubing, laminar flow in the tubing will prevent the measurement of abrupt concentration changes. This condition will exist in oceanographic studies when samples from great depths are to be analyzed or where the analyzer must be located at some distance from the stream. In these cases a separate high velocity pump should be used to provide the water to a small reservoir from which the Auto-Analyzer pump can withdraw a sample. A sampler for oceanographic use which segments the sample with air at the sampling point has been described.[2] Armstrong and LaFond[1] show the results of continuous silicate and nitrate analyses off the coast of California.

Frequently continuous analyses for several constituents are desired. Large numbers of AutoAnalyzers are not only expensive but require a large amount of space to operate. The Continuous Simultaneous Monitor CSM 6 is an instrument designed for the measurement of six components. Continuous water sampling and clarification are shown in Figure 23. Its proportioning pump operates at a slower speed than those for the regular AutoAnalyzer, thus permitting unattended operation for a long time. A single colorimeter containing a flow cell and photocell for each determination and one reference photocell are used. The CSM 6 provides, in addition to results presented on strip-chart or multipoint recorders, a 0-5 volt DC output with variable expansion for each channel. This signal is compatible with most punched paper or magnetic tape data acquisition systems. Discrete samples can be analyzed using a sampler module. Some parameters which have been measured with this system are listed in Table 17.

Data Collection

Continuous monitoring systems are generally used with multipoint or strip-chart recorders to record the data. These data often must be transcribed to permit further analysis such as the computation of the average. In many applications, it is necessary to obtain the information immediately so that remedial action can be taken. The sheer bulk of data which can be obtained from a monitoring system requires the use of supplemental means of data acquisition. An eight parameter system, for example, generates 5376 data points per week if each channel is monitored on a fifteen minute cycle.

118

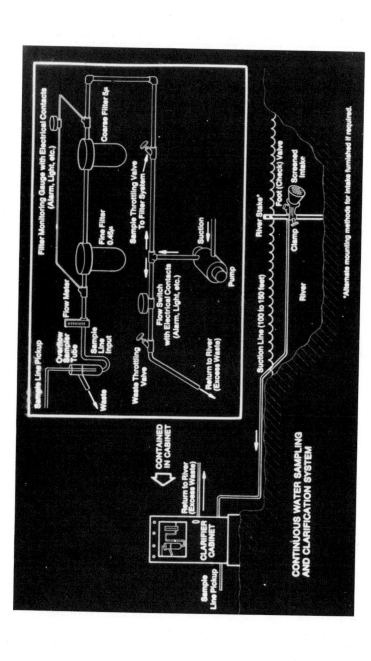

Figure 23. Continuous water sampling and clarification system.

Table 17

Parameters Measured by the Technicon CSM 6*

Parameter	Nominal(a) Range (ppm)	Detection Limit (ppm)
Phosphate	0–8	0.08
Chromium (hexavalent)	0–5	0.05
Copper	0–10	0.10
Iron	0–10	0.10
Ammonia	0–10	0.10
Methyl Orange alkalinity(b)	0–500	5.0
Thymol Blue alkalinity(b)	0–100	1.0
Hardness(b)	0–300	3.0
Sulfate(c)	0–500	5.0
Phenol	0–5	0.05
Cyanide(b)	0–3	0.03
Chemical oxygen demand	0–500	5.0
Chloride(b)	0–10	0.10
Nitrate	0–1.5	0.015
Nitrite + Nitrate	0–2.0	0.02
Fluoride(b)	0–2.5	0.025
Orthophosphate	0–10	0.10
Silicate	0–15	0.15

* From Reference 11.
(a) Alternate ranges optional.
(b) Deviates from linearity.
(c) Sulfate is only available on continuous monitoring applications.

Supplemental data acquisition can be local or the data can be telemetered to a remote location. Telemetering of either digital or analog data is possible. Analog transmission, accomplished by variation of the frequency of the signal, requires less expensive equipment than that for digital transmission, but there is a loss in accuracy during transmission of the data. The transmitted data may be recorded in analog for another multipoint recorder, converted to digital form for recording on punched paper or magnetic tape to be used by a computer, or automatically typed on a log sheet. For digital transmission an analog

to digital converter sequentially converts the measurement
signals to pulsed digital information which is transmitted
by telephone to a central station. This data may be recorded on tape for later computer input or may be directly
fed into a computer. When storing data on tape for later
print-out or computer processing, real time should be
recorded from a digital clock to permit complete identification of samples at a later time. These and other means
of data acquisition and telemetry are shown in Figure 24.

The use of continuous monitoring equipment will permit
the frequent measurement of several water quality parameters and the recording of this data at the measurement
site or the transmitting of it to another location. The
choice of the best method of data recording and of sensors
to be used depends on the purpose of the monitoring. If
the monitoring is to provide information on fluctuations
in quality which will be used as background data for
future studies, telemetering of the data is usually
unnecessary. In most studies, however, telemetering of
data is important because the scientist may wish to begin
sampling for other constituents when one of the continuously measured parameters reaches some high or low value.
Water management decisions, such as stopping waste effluent from entering a river or temporary suspension of
drawing water from a river, may result from decisions
based on these telemetered data. To call attention to
values which fall outside the present level, automatic
alarms may be added to the system and samples can be
collected automatically by a similar circuit actuating an
automatic sampling system.

Sensors

It would be desirable to have at our disposal a series
of sensors, each specific to one physical or chemical
parameter of water quality. This level of sophistication
has not been achieved, but many analyses can be made continuously or on a sequential basis by automated equipment.
The input transducer or sensor responds to a physical or
chemical change by a proportional change in a transmittable
signal, most commonly an electrical signal. This signal
is modified in the measuring system, which can be considered as a series of transducers:[10] (a) input or measuring transducers, (b) modifying transducers, and (c) output or readout transducers.

Temperature sensors have been recently reviewed by
Bollinger,[4] who indicates that resistance thermometers,
thermistors, and thermocouples may be used for continuous

121

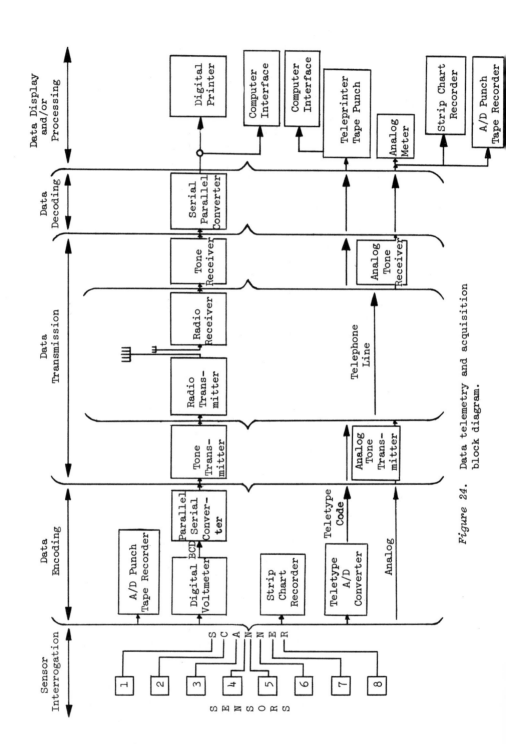

Figure 24. Data telemetry and acquisition block diagram.

monitoring. Resistance thermometers are based on the temperature dependence of the resistance of most metals. From accurate measurement of the resistance of a coil of metal wire, often nickel or platinum, the temperature can be determined. Thermistors are semiconductors showing a high negative coefficient of resistance as a function of temperature. Changes of resistance by a ratio of 10:1 over the range -100 to +450°C are known. In thermocouples a potential proportional to the difference between the sample and reference temperatures is developed at the junction of two dissimilar metals (such as Chromel-Alumel).

A number of sensors which are suitable for continuous monitoring of chemical properties are listed in Table 18. Many of these have been used in continuous monitoring applications.[3] Generally the most serious limitation to the use of many of these sensors is their need to be standardized frequently. This requirement is a serious limitation to their use in remote locations where it may be desirable to service an instrument weekly or semiweekly.

Table 18

Continuous Sensors

Electrical Sensors

    Amperometric
    Conductometric (electrical)
    Coulometric
    Galvanic
    Ionization detectors
    Polarographic
    Potentiometric

Optical Sensors

    Colorimetric (psectrophoto-
      metric, visible region)
    Emission spectroscopic
    Flame photometric
    Fluorometric
    Infrared spectrophotometric
    Particle counting
    Turbidimetric
    Ultraviolet spectrophoto-
      metric

Sensors Based on Physical Properties

    Density
    Optical rotation
    Refractive index
    Viscosity

Sensors Based on Thermal Properties

    Conductivity (thermal)

(cont)

Table 18

(cont)

Sensors Based on Miscellaneous Properties	Sensors Based on Radioactivity
Magnetic Mass spectroscopy	

*From Reference 3.

At this time many new specific ion electrodes (Table 19) have become commercially available. These electrodes can be adapted easily for use with existing continuous systems by including additional signal conditioning modules for the measurement of electrode potential.

The simplest continuous monitoring system would consist of a specific ion electrode and a suitable reference electrode placed in a sample stream. The output of the pH meter can be recorded using a suitable strip-chart recorder. Such systems have been used for the recording of pH and, with suitable controls and pumps, can be used to add acid or base to maintain a desired pH. Similar systems can be developed using other specific ion electrodes as sensors.

When designing monitoring systems using specific ion electrodes, the analyst should keep clearly in mind the factors controlling the electrode response and the solution parameter which is desired. Since electrodes measure the activity of the ion, their response will vary with the ionic strength of the solution. In addition metals such as cadmium, lead or copper complex with hydroxide, and anions such as fluoride and cyanide react with hydrogen ions. Often the total concentration is sought rather than the activity of the species directly measured by the electrode. Addition of a constant flow of a buffer containing a high concentration of an inert electrolyte is one method of obtaining the desired results. The effect of pH on the species being measured can be computed as illustrated by Equations 61-63 for the case of sulfide. By measurement of both pH and sulfide in the unaltered sample stream the total concentration of sulfide as well as the concentration of each species ($S^=$, HS^-, and H_2S)

Table 19

Commercially Available
Ion-Selective Electrodes

Ion	Membrane	Lower Detection Limit	Principal Interferences
H^+	glass	10^{-14} M	none below pH 13
Na^+	glass	10^{-6} M	H^+, Ag^+
K^+	glass	10^{-6} M	H^+, Ag^+, Na^+
NO_3^-	liquid	10^{-5} M	I^-, Br^-
Cl^-	liquid	10^{-5} M	I^-, NO_3^-, Br^-
Ca^{++}	liquid	10^{-5} M	Zn^{++}, Fe^{++}, Pb^{++} Cu^{++}
Water hardness	liquid	10^{-8} M	Zn^{++}, Fe^{++}, Pb^{++}
F^-	solid	10^{-6} M	OH^-
Cl^-	solid	5×10^{-5} M	$S^=$, I^-, Br^-, CN^-
Br^-	solid	5×10^{-6} M	I^-, $S^=$
I^-	solid	5×10^{-8} M	$S^=$
$S^=$	solid	10^{-17} M	Hg^{++}
CN^-	solid	10^{-6} M	$S^=$, I^-
Ag^+	solid	10^{-17} M	Hg^{++}
Cd^{++}	solid	10^{-7} M	Ag^+, Cu^{++}, Hg^{++}
Pb^{++}	solid	10^{-7} M	Ag^+, Cu^{++}, Hg^{++}
Cu^{++}	solid	10^{-8} M	Ag^+, Hg^{++}, $S^=$

may be computed. Such computations may be made in real time by analog computation methods.[5]

The results of electrode measurements will be low if the sample contains complexing agents because the complexed species will not be detected by the electrode. Such is the case in the measurement of fluoride where aluminum and iron in the sample complex with fluoride. A buffer containing citrate has been used to release fluoride from its complex with aluminum or iron as well as to control the pH and ionic strength.[6]

Several possible measurement schemes can be used if the reagent which is added to the sample stream contains the ion being measured or a complexing agent which can be measured by a specific ion electrode.[8] In all cases two identical specific ion electrodes are used in the measurement circuit, eliminating the need to use conventional reference electrodes with a liquid junction. The two electrodes can be inserted in three locations as shown in Figure 25. These locations are in the reagent line (R), the sample line (S), and the line containing the sample and reagent or the total line (T).

If the reagent contains the ion being measured, the potential difference $(E_T - E_S)$ gives the concentration of the ion by the method of known addition. The potential difference between electrodes in the reagent and total lines $(E_R - E_T)$ is treated by the method of analate addition. The potential difference $(E_R - E_S)$ is a direct measurement of the activity of the ion in the sample stream, since the potential of the electrode in the reagent line is constant. If the reagent stream contains a complexing agent, the potential difference $(E_T - E_S)$ can be used to calculate the concentration of the ion in the sample by the method of known subtraction.

Ions for which there is no electrode can be measured if they complex or precipitate with an ion which can be measured. By using lead perchlorate as a reagent the potential difference $(E_T - E_R)$ between a pair of lead-specific ion electrodes and the data is treated by the method of analate subtraction.[8]

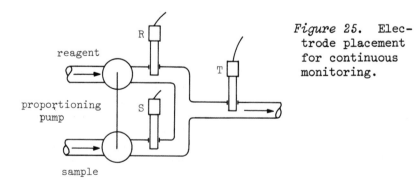

reagent

proportioning
pump

sample

Figure 25. Electrode placement for continuous monitoring.

Mathematical treatments of both the case of known addition or subtraction[7] and analate addition or subtraction[8] have been developed. For known addition or subtraction the expression for the change in potential is

$$\Delta E = \frac{RT}{nF} \log \left[\frac{\gamma_2 f_2 (v_s c_s \pm v_r c_r)}{\gamma_1 f_1 c_s (v_s + v_r)} \right] \tag{28}$$

For analate addition or subtraction the expression for the change in potential is

$$\Delta E = \frac{RT}{nF} \log \left[\frac{\gamma_2 f_2}{\gamma_1 f_1 (v_r + v_s)} \left(v_r \pm \frac{v_s c_s}{c_r} \right) \right] \tag{29}$$

For both expressions the positive sign is for additions and the negative sign is for subtractions. The terms in the above expressions are defined:

v_r = the volume of reagent solution

v_s = the sample volume

c_r = the concentration of the ion sensed by the electrode in the reagent solution

c_s = the concentration of the ion in the sample solution

γ_1, γ_2 = the activity coefficient of the ion to be determined before and after combining sample and reagent

f_1, f_2 = the fractional complexation before and after combining sample and reagent

ΔE = the change in potential obtained by mixing the sample and reagent

$\frac{RT}{F}$ = the Nernst factor

n = the charge of the ion sensed by the electrode.

Analysis and Control of Biological Processes

As the necessary analytical instrumentation becomes available, waste treatment plants will be able to substitute these methods rather than use wet chemical methods. These modern methods could be performed on a continuous basis not only to provide instantaneous information on the efficiency of the plant but also to permit automatic control of the waste treatment facility. This modernization will result in optimization of the treatment process which will insure that the effluent will be of high quality over a wide range of operating conditions. Automation will also reduce the manpower requirements of modern treatment facilities.

127

Zenz, McAloon and Weddle[12] have reviewed many of the applicable analyzers which could be used for continuous monitoring for the activated sludge process and anaerobic digestion. Some of the parameters which they considered it essential to monitor continuously included substrate concentration, suspended solids, toxicants, pH, temperature, dissolved oxygen, nitrogen, phosphorus, conductivity, biological activity, gas composition and residual chlorine.

A knowledge of the substrate concentration is essential to optimization of treatment. The time lag in the measurement of BOD is too great to permit this determination to be used in controlling plant operations. The introduction of rapid instrumental methods permits the measurement of substrate concentrations on a time scale required for the obtaining of operating information.

Chemical Oxygen Demand (COD) can be measured using the Technicon AutoAnalyzer as indicated in Chapter VII. The Precision Aquarator (Precision Scientific Co.) measures COD in approximately two minutes by using dry CO_2 to carry the organic matter through a platinum catalytic combustion furnace which oxidizes it to CO and H_2O. Following water removal and passage through a second catalytic treatment, the CO concentration is measured using an infrared analyzer (Figure 26).

In Total Carbon Analyzers (Beckman Instruments, Inc. and Union Carbide Corp.) organic matter is combusted to CO_2, H_2O, and N_2. The CO_2 concentration is determined after separation of the H_2O by an infrared analyzer (Figure 27). The analysis will determine the total carbon content of the sample unless the carbonate is removed by acidification and nitrogen purging of the sample before analysis. If this is done, the sample can be analyzed for total organic carbon (TOC).

Another analyzer (Rocketdyne) employs pyrolytic fragmentation of the organics in the presence of water. The fragments are then measured with a hydrogen flame ionization detector. This analyzer does not respond to inorganic carbonates or dissolved carbon dioxide in the sample.

Suspended solids can be continuously measured by the absorption or reflection of energy. At present, visible light, infrared radiation, ultrasonic waves and gamma radiation are used for solids determinations. Of these, visible light has been used the most. Because of its limited penetrating power, its absorption or reflection depends mainly on the surface area of the solids rather than their volume and density. These instruments require low flows to prevent light losses at the liquid surface.

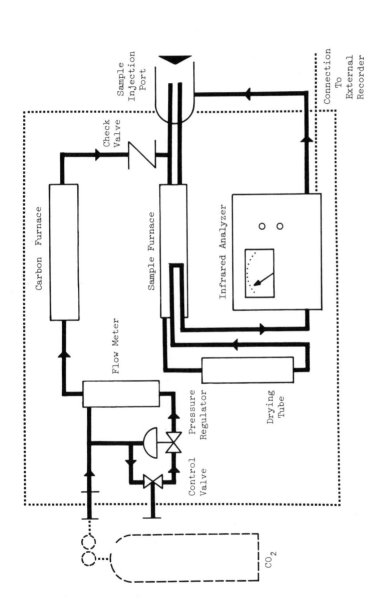

Figure 26. COD Analyzer flow diagram.

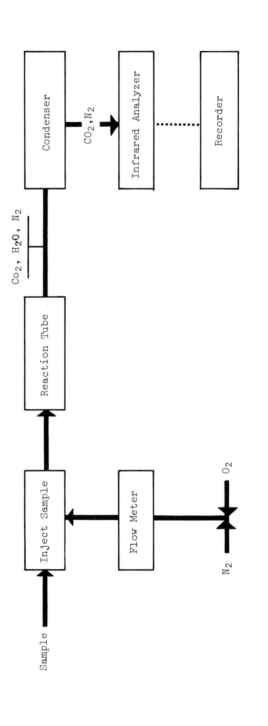

Figure 27. Total Carbon Analyzer flow diagram.

In waste waters of high suspended solids, clogging may occur.

Analyzers employing infrared radiation measure dissolved solids concentrations more satisfactorily than visible light instruments due to the greater penetrating power of infrared radiation. Ultrasonic radiation does not appear to be sensitive enough to measure suspended solids concentrations of less than one per cent. The equipment required can be as much as ten times as expensive as visible or infrared measuring devices.

Gamma radiation devices using cesium-137 or other isotopes lacking strong beta radiation have great penetrating power. This equipment is applicable to monitoring large process lines for the solids concentrations found in the activated sludge process. The equipment required is expensive, and it requires special maintenance, calibration and supervision.

The biological activity in the activated sludge process has not been monitored satisfactorily on a continuous basis, but several new methods appear to be applicable to the new problem. The activity of the dehydrogenase enzyme can be monitored continuously by a colorimetric analytical procedure and can be used to control the process. The reaction associated with the measurements is as follows:

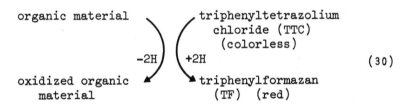

$$\begin{array}{cc}
\text{organic material} & \text{triphenyltetrazolium} \\
 & \text{chloride (TTC)} \\
 & \text{(colorless)} \\
-2H \qquad +2H & \qquad\qquad (30) \\
\text{oxidized organic} & \text{triphenylformazan} \\
\text{material} & \text{(TF) (red)}
\end{array}$$

The TF is insoluble in water but soluble in alcohol.

Biological activity could also be monitored by measuring the rate of oxygen uptake. Samples can be withdrawn automatically and their oxygen concentration monitored for a given time using a dissolved oxygen probe.

The active mass of bacteria may be measured not only in waste reactors but also in natural water systems by monitoring the amount of adenosine triphosphate (ATP). This procedure involves adding the sample to a buffered reaction mixture of luciferin and luciferase (materials obtained from firefly tails) and measuring the light produced.

Two basic types of instruments are available for the measurement of conductance. The more familiar system imposes an AC potential between a pair of platinum electrodes and measures their resistance which is a function of the conductance of the solution. Electrodeless conductivity systems eliminate the problem of coating the electrodes. The system measures changes in inductance or capacitance due to the sample which is not in physical contact with the cell. The inductance or capacitance can then be related to the sample's conductance.

Residual chlorine can be monitored by amperometric instrumentation. The sample passes a pair of electrodes of dissimilar metals. The polarization of the nobler metal electrode prevents the passage of current in the absence of an oxidizing agent. Traces of oxidizing agents cause depolarization, and a current proportional to the concentration of oxidants is generated if a constant potential is maintained across the electrodes. Since a continuous signal is obtained from this type of instrument, it is possible to automatically control the addition of chlorine to the system.

The gas composition and volatile acids content of anaerobic digestors can be monitored by gas chromatography. The automation or gas chromatography has been achieved for some industrial applications, but this has not yet been extended to the waste treatment field. Further research and development will be required before gas chromatography can be used to continuously monitor and control the anaerobic digestion process.

Many of the other instrumental methods required for the monitoring of waste treatment facilities are discussed elsewhere in this book. Nitrogen, phosphorus and specific toxicants can be measured by the AutoAnalyzer while heavy metals can be analyzed by atomic spectroscopy or with specific ion electrodes for some metals. Electrode systems are also used for the measurement of dissolved oxygen and pH while temperature can be measured by thermistors or thermocouples.

REFERENCES

1. Armstrong, F.A.J., and E.C. LaFond. Limnol. Oceanogr. 11, 538 (1966).
2. Bernhard, M., and G. Macchi. *Automation in Analytical Chemistry; Technicon Symposia 1965* (New York: Mediad, 1966) p 255.
3. Blaedel, W.J., and R.H. Laessig. *Advances in Analytical Chemistry and Instrumentation*, Vol. 5, C.N. Reilley and F.W. McLafferty, Eds., (New York: Interscience Publishers, 1966) p. 69.

4. L. Bollinger. *Environmental Measurements: Valid Data and Logical Interpretation*, U.S. Public Health Service Publ. 999-WP-15, Washington, D.C. (1964) p 41.
5. Dimirjian, Y., and K.H. Mancy. Symposium on the Applications of Selective Ion Electrodes, Pittsburgh Conference on Analytical Chemistry and Applied Spectroscopy, Cleveland, Ohio (April, 1969).
6. Frant, M.S., and J.W. Ross, Jr. Anal. Chem. 40, 1169 (1968).
7. Orion Research Incorporated, Newsletter/Specific Ion Electrode Technology, Cambridge, Mass. 2, 5 (1970).
8. Orion Research Incorporated, Newsletter/Specific Ion Electrode Technology, Cambridge, Mass. 2, 21 (1970).
9. Porterfield, H.W. Oceanology International 5 (10), 22 (1970).
10. Stein, P.K. *Environmental Measurements: Valid Data and Logical Interpretation*, U.S. Public Health Service Publ. 999-WP-15, Washington, D.C. (1964) p 65.
11. Technicon Corp., Environmental Sciences Division, Tarrytown, N.Y. (1969).
12. Zenz, D.R., T.J. McAloon, and C.L. Weddle. *Sensors for the Analysis and Control of Biological Processes*, Division of Water, Air, and Waste Chemistry, American Chemical Society, Minneapolis, Minn. (April, 1969).

Chapter VII

Automated Chemical Analyses

The water analyst is now faced with a rapidly increasing number of requests for routine analyses of water
samples. To facilitate the analysis of large numbers of
samples, the chemist has begun using automated methods of
analysis. It is the purpose of this chapter to describe
the Technicon AutoAnalyzer, an instrument for automated
colorimetric chemical analyses, and to detail important
points in its operation.

Automation of wet chemical methods not only permits
analysis of more samples than can be analyzed manually,
but also offers greater precision than manual methods.
Human errors are minimized by the automatic proportioning
of sample and reagent and development and measurement of
the color produced. Since the entire procedure, including
reagent addition, mixing and heating, is exactly repro-
duced, methods which might be difficult or impossible
manually can be used on the AutoAnalyzer. However, when
an automated procedure is adopted, the chemist loses much
of the control over specific steps that he had with manual
methods. Due to the time required to prepare the Auto-
Analyzer for an analysis, it is not advantageous to run
fewer than twenty tests except for the most difficult
manual procedures.

System Design

The AutoAnalyzer is an instrument capable of automatic
sampling, filtering, diluting, reagent addition, mixing,
heating, digesting, and, after appropriate delay for color
development, measurement of the color produced (Figure 28).

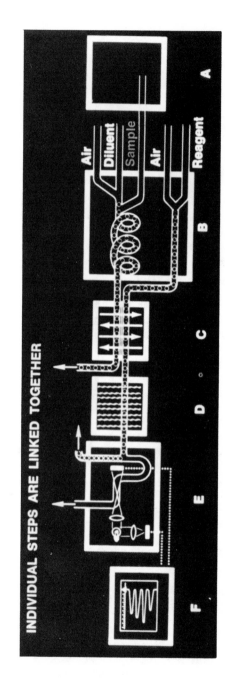

Figure 28. Linking of individual steps in automated analysis: (A) sample supply, (B) pump for the addition of reagents, mixing, and air segmentation, (C) continuous dialyzer, (D) heating bath, (E) colorimeter with de-bubbler for removal of air prior to measurement, and (F) a recorder.

All operations are performed on a stream which is moved by a fixed speed peristaltic pump. The instrument is modular and reconfiguration of the modules for different analyses takes only a few minutes.

The sampler contains a removable circular turntable which holds 40 sample cups. Samples and standards, whose volumes do not have to be measured, are placed in the cups. Between samples the probe aspirates a wash solution from a receptacle which is kept filled by the pump. A cam determines the aspiration time of the sample and wash solution. The ratio of sample to wash time is usually 2:1 and 20-60 tests/hour are performed, although cams are available for 10-120 tests/hour with 11 different sample-to-wash ratios of 6:1 to 1:6. The sampling cycle is (1) the probe dips into the sample cup, (2) the sample is aspirated, (3) the probe withdraws from the sample cup and dips into the wash receptacle, and (4) the wash solution (generally water) is aspirated.

A sampler which holds 100 test tubes is also available. This sampler is useful in water pollution studies since samples can be collected in the same container that is placed in the sampler. There is usually enough sample contained in one test tube that several simultaneous analyses for different parameters can be measured on a single sample.

The proportioning pump is the heart of the analytical system. It draws the sample, reagents, and air, which segments the sample, at a fixed rate through individual tubes. The pump operates by peristaltic action as shown in Figure 29. Rollers occlude the liquid contained in the taut tubing and force it forward at a fixed linear velocity with a liquid flow rate proportional to the inside diameter of the tubing. Twenty different tubing diameters from 0.005 to 0.110 inches providing flows of 0.015 to 3.90 ml/min are available (Table 20).

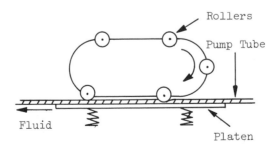

Figure 29. Proportioning pump. Rollers occluding pump tubes on platen.

137

Table 20

Tube Sizes and Deliveries with Standard AutoAnalyzer Pump

Tube		Clear Standard Delivery ml/min			Solvaflex Delivery ml/min			Acidflex Delivery ml/min (approx)
I.D. (inches)	Shoulder Colors	Min.	Nom.	Max.	Min.	Nom.	Max.	
.005	orange black	0.005	0.015	0.029	0.005	0.015	0.029	n.a.*
.0075	orange red	0.016	0.03	0.048	0.016	0.03	0.048	n.a.
.010	orange blue	0.032	0.05	0.072	0.032	0.05	0.072	n.a.
.015	orange green	0.075	0.10	0.128	0.075	0.10	0.128	n.a.
.020	orange yellow	0.13	0.16	0.19	0.13	0.16	0.19	n.a.
.025	orange white	0.19	0.23	0.27	0.19	0.23	0.27	n.a.
.030	black	0.28	0.32	0.36	0.28	0.32	0.36	0.34
.035	orange	0.37	0.42	0.47	0.37	0.42	0.47	0.43
.040	white	0.54	0.60	0.66	0.51	0.56	0.62	0.53
.045	red	0.73	0.80	0.87	0.64	0.70	0.76	0.64
.051	grey	0.92	1.00	1.03	0.81	0.88	0.93	0.78
.056	yellow	1.12	1.20	1.28	0.99	1.06	1.13	0.92
.060	yellow blue	1.31	1.40	1.49	1.14	1.21	1.29	1.06
.065	blue	1.50	1.60	1.70	1.29	1.37	1.45	1.19
.073	green	1.90	2.00	2.10	1.60	1.69	1.78	1.44
.081	purple	2.37	2.50	2.63	1.92	2.02	2.12	1.71
.090	purple black	2.77	2.90	3.03	2.31	2.42	2.53	2.03
.100	purple orange	3.26	3.40	3.54	2.77	2.89	3.01	2.39
.110	purple white	3.75	3.90	4.05	3.26	3.39	3.52	2.76

*n.a. = not available

Pump tubes of three materials are available: standard
clear plastic, solvaflex yellow plastic, and acidflex, a
black plastic. Liquid compatibility with the various types
of tubing is indicated in Table 21. Liquids which are
incompatible with any of the types of tubing may be pumped
using a displacement bottle. Water or another immiscible
liquid is pumped into a closed bottle containing the incompatible liquid which is then carried by glass or other
inert tubing to a fitting where it will be diluted by
sample or reagent.

Table 21

Recommended Tubing for Specific Chemicals

Chemical	Tubing
Acetaldehyde, dilute	clear standard
Acetic acid 95%, water 5%	acidflex
Acetone	displacement bottle
Acids, mineral (except hydrofluoric	
dilute	clear standard or acidflex
concentrated	acidflex
Amyl acetate	displacement bottle
Aqueous solutions	clear standard
Benzene	acidflex
Carbon dioxide (gas)	clear standard
Carbon tetrachloride	solvaflex
Chloroform	acidflex
Dioxane	displacement bottle
Ethyl alcohol	solvaflex
Formaldehyde	clear standard
Gasoline	solvaflex
Glycerine	clear standard
Glycol	clear standard
Isopropyl alcohol	solvaflex
Methyl alcohol	solvaflex
Methyl cellosolve, 65%	solvaflex
Octane	solvaflex
Pyridine	displacement bottle
Sodium hydroxide	clear standard
Styrene	acidflex
Toluene	acidflex
Trichloroacetic acid	acidflex
Water	standard
Xylene	acidflex

The pump can accommodate 15 or 23 tubes depending on
the model. The tubes containing sample, air and various
reagent streams are brought together by 'h', 'T' or other
shaped glass fittings. The liquid stream is segmented by
air bubbles to prevent diffusion or cross-contamination of
samples due to laminar flow in the narrow tubing. Air
segmentation greatly reduces the time necessary to achieve
steady-state readings and is responsible for the great
speed of analyses. Without air segmentation, the wash
time between samples is greater than five minutes for many
procedures while with air segmentation rarely is more than
one minute necessary.

Each reagent added requires a pump tube of appropriate
diameter which is joined to the main stream by a glass
fitting. After the addition of a reagent, the stream must
be mixed by being passed through a glass coil as shown in
Figure 30. The mixing is achieved in a coil whose axis
is horizontal, causing the heavier liquid to fall through
the lighter liquid. Mixing coils with 7, 14 and 28 turns
are available.

Many types of mixing coils are available. Regular
tubes (2.4 mm I.D.) are normally used, but with high flow
rates 3.4 mm I.D. coils may be needed for effective mixing.
With the large bore tubing, air segmentation of the liquid
must be from a 0.065 inch I.D. or larger pump tube. Jacketed mixing coils to cool streams emerging from the
heating bath and coils with reagent inlets at the end or
center of the coil are available. Jacketed glass fittings
are also available and are used when adding concentrated
acids to water.

Liquid-liquid extractions may be performed on the Auto-
Analyzer, but it is, of course, impossible to use a high
ratio of aqueous to organic phases. Extraction efficiency

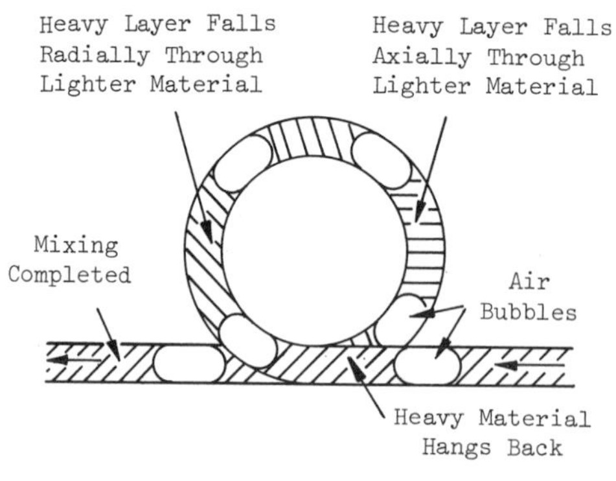

Heavy Layer Falls
Radially Through
Lighter Material

Heavy Layer Falls
Axially Through
Lighter Material

Mixing
Completed

Air
Bubbles

Heavy Material
Hangs Back

Figure 30.
Mixing coil –
fluid inversion
cycle.

can be improved by using mixing coils containing glass beads. These coils are available in single and double length, with or without water jackets. Phases may be separated using the phase separator described by Hadley[30] or a glass fitting used for debubbling a sample stream.

The system of pump tubes, their supports which keep the tubes taut, mixing coils, glass fitting and connecting tubing is called a manifold. Each analysis requires its own distinctive manifold. A plastic platter which fits snugly onto the pump is used to support the manifold and to hold its components rigid. The platter permits rapid changing of manifolds, so one AutoAnalyzer may be used for several different analyses each day.

Time-delay coils and heating baths provide the necessary time and heat for reactions to take place. The coils are horizontal and therefore provide no mixing. Coils 10, 20, and 40 feet long are available. Heating baths used for water analyses usually are provided with adjustable thermoregulators with maximum temperatures of 100° or 200°C. The higher temperature is needed for the determination of chemical oxygen demand and in procedures employing a distillation (*e.g.*, phenol and fluoride).

The continuous digester module has been used primarily for the automation of Kjeldahl nitrogen analyses, but has also found use in the digestion of organic phosphates[7] and in the analysis of chemical oxygen demand.[70] The digester is capable of operating at temperatures as high as 700°C; for any procedure the digester may be operated at a high enough temperature to insure rapid and complete destruction of organic matter. Sample and digestion fluid are pumped into the glass digester tube which is helically grooved and is slowly rotated by a motor, causing the sample to be spirally transported to the other end. Two independently controlled heaters are used in the digester. The first stage heater causes the water to be vaporized and begins the oxidation of the sample. The second stage heater, which is maintained at a higher temperature than the first, completes the oxidation of the sample. At the exit end of the helix, water is added and the diluted material is aspirated into a small mixing chamber before passing to the manifold where the chromogenic reagents will be added to the sample.

The color measurement is performed in a filter photometer equipped with a flow-cell. Before the sample stream enters the colorimeter, air bubbles are removed. The debubbling is achieved by pulling only a portion of the liquid stream through the colorimeter cell by a tube in the pump. Air bubbles and the remaining portion of the

liquid stream are discharged to a waste line. Figure 31
shows the details of the flow cell and debubbler assembly.
Flow cells of 8- and 15-mm path length can be accommodated
in the colorimeter, but a 50-mm cell is generally used for
water analyses in order to gain sensitivity. Range ex-
panders are often required to improve the readability of
the recorder chart. The range expander allows full scale
recorder deflection to represent 50-100% (2x), 75-100%
(4x), or 90-100% transmission (10x). With the adjustable
model range expander any 50%, 25%, or 10% range may be
magnified to full scale deflection. For many tests on
natural water the analyst deals with concentrations near
the limit of sensitivity of available analytical procedures and must use both long cells and a range expander.

The per cent transmission of the samples is recorded on
the recorder chart. For those tests obeying the Beer-
Lambert law, the per cent transmission of the standards
can be plotted on semilogarithmic paper if no range expansion was used. If 4x or 10x range expansion was used, the
recorder response should be plotted on linear paper since
in the ranges of 75-100 and 90-100 the logarithm of a
number and the number are approximately proportional.
When 2x range expansion is used, the plot will generally
be most nearly linear if made on semilogarithmic paper.
For the most precise work in all cases, absorbances should
be calculated and plotted vs. the concentration. Table 22
is a listing of a Fortran IV program which computes absorbances of samples from the recorder response. The
slope and intercept of the standard series is computed by
the method of least squares and this data is used to com-
pute the absorbance of the samples. Drift of the base
line is assumed to be linear and is compensated for.

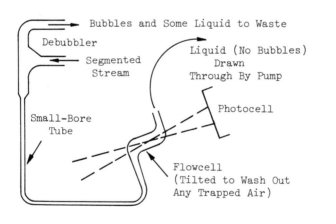

Figure 31.
Flowcell
and de-
bubbler.

Table 22

Computer Program (Fortran IV) for the Analyses of AutoAnalyzer Data

```
C........ LIST OF VARIABLES........
C...
C...A                        SUM OF ABSORBANCES
C...C                        SUM OF CONCENTRATIONS
C...AA                       SUM OF (ABSORBANCES)**2
C...CC                       SUM OF (CONCENTRATIONS)**2
C...AC                       SUM OF (ABSORBANCE)*(CONCENTRATION)
C...CONST                    CONVERSION CONSTANT
C...NSTD                     NUMBER OF STANDARDS
C...NSAMP                    NUMBER OF SAMPLES
C...RANGE,IRANG              RANGE EXPANSION
C...TBASB                    BASE LINE TRANSMITTANCE BEFORE ANALYSES
C...TBASA                    BASE LINE TRANSMITTANCE AFTER ANALYSES
C...MO                       MONTH
C...DAY                      DAY
C...COMM                     COMMENT FIELD
C...ABASB                    BASE LINE ABSORBANCE BEFORE SET
C...ABASA                    BASE LINE ABSORBANCE BEFORE SET
C...DBASE                    DIFFERENCE IN BASE LINE BEFORE AND AFTER (DRIFT)
C...CONC                     CONCENTRATION OF STANDARDS
C...TRANS                    TRANSMITTANCE OF PERCENT TRANSMISSION*100.
C...COM                      COMMENT FIELD
C...ABSM                     MEASURED ABSORBANCE
C...ABS                      TRUE ABSORBANCE
C...SLOPE                    SLOPE OF CONCENTRATION VS. ABSORBANCE
C...INCPT                    INTERCEPT OF ABSORBANCE AXIS
C...CORCO                    CORRELATION COEFFICIENT
C...NUM                      NUMBER OF STANDARDS + SAMPLES + 1
C...IBOT                     SAMPLE IDENTIFICATION (BOTTLE NUMBER)
C...
      INTEGER DAY
      REAL INCPT
      DIMENSION COMM(14),COM(7)
      CONST=0.4342945
C........INITIALIZATION OF VALUES USED FOR LEAST SQUARES ANALYSIS
   10 A=0.
      C=0.
      CC=0.
      AA=0.
      AC=0.
C........INPUT FOR SET CONTROL CARD
C........VARIABLE.....................INPUT COLUMNS....FIELD TYPE....EXAMPLE..
C...
C...NSTD    (NUMBER OF STANDARDS)            1-3          I3          006
C...NSAMP   (NUMBER OF SAMPLES)              4-6          I3          012
C...RANGE   (RANGE EXPANSION)                7-8          F2.0         01
C...TPASB   (BASE LINE PERCENT TRANSMISSION  9-12         F4.3        1000
C...TBASA    *100. BEFORE AND AFTER SET)    13-16         F4.3        0975
C...MO      (MONTH)                         17-18         I2           02
C...DAY     (DAY)                           19-20         I2           14
C...COMM    (COMMENT FIELD)                 21-76         14A4      **TITLE**
C...
      READ(2,20)NSTD,NSAMP,RANGE,TBASB,TBASA,MO,DAY,COMM
   20 FORMAT(2I3,F2.0,2F4.3,2I2,14A4)
      IF(NSTD) 100,100,27
C........BASE LINE DRIFT COMPUTED (DBASE)
   27 ABASB=CONST* ALOG(1./(1.-(1.-TBASB)/RANGE))
      ABASA=CONST* ALOG(1./(1.-(1.-TBASA)/RANGE))
      DBASE=ABASA-ABASB
C........HEADING FOR OUTPUT OF STANDARDS
      WRITE(3,11)MO,DAY
```

(cont)

Table 22
(cont)

```
   11 FORMAT(1H1,50X,I2,'/',I2)
      WRITE(3,14)COMM
   14 FORMAT(1H0,30X,15A4// 50X,'STANDARDS'/)
      WRITE(3,15)
   15 FORMAT(1H ,T32,'MEASURED',T47,'MEASURED',T63,'TRUE'/ T30,'TRANSMIS
     1SION ',T46,'ABSORBANCE',T60,'ABSORBANCE',T75,'CONCENTRATION'/)
C........INPUT FOR STANDARDS
C........VARIABLE......................INPUT COLUMNS....FIELD TYPE....EXAMPLE..
C...
C...CONC   (CONCENTRATION OF STANDARD)       1-6            F6.2        025000
C...TRANS  (PERCENT TRANSMISSION * 100.)     7-10           F4.3          0750
C...COM    (COMMENT FIELD)                   11-38          7A4       **REMARKS*
C...
      DO 40 I=1,NSTD
      READ(2,50)CONC,TRANS,COM
   50 FORMAT(F6.2,F4.3,7A4)
C........COMPUTE MEASURED AND  TRUE ABSORBANCE
      ABSM=CONST* ALOG(1./(1.-(1.-TRANS)/RANGE))
      NUM=NSTD+NSAMP+1
      ABS=ABSM-ABASB-(I*DBASE/NUM)
C........OUTPUT FOR STANDARDS
      WRITE(3,60)TRANS,ABSM,ABS,CONC,COM
   60 FORMAT(1H ,T33,F6.4,T47,F7.4,T61,F7.4,T78,F8.3,T89,7A4)
C........COLLECT DATA FOR LEAST SQUARES ANALYSIS
      A=A+ABS
      C=C+CONC
      AA=AA+ABS*ABS
      CC=CC+CONC*CONC
   40 AC=AC+ABS*CONC
C........COMPLETE THE LEAST SQUARES ANALYSIS AFTER INPUT OF ALL STANDARDS
      SLOPE=(NSTD*AC-A*C)/(NSTD*CC-C*C)
      INCPT=A/NSTD-SLOPE*C/NSTD
      CORCO=(NSTD*AC-C*A)/SQRT((NSTD*CC-C*C)*(NSTD*AA-A*A))
      IRANG=RANGE
      WRITE(3,70)SLOPE,INCPT,CORCO,IRANG,ABASB,ABASA
   70 FORMAT(1H0,5X,110('*')/ 10X,'SLOPE = ',E12.4,10X,'INTERCEPT = ',
     1F7.4/10X,'CORRELATION COEFFICIENT = ',F6.4,10X,'RANGE EXPANSION '
     2I2,'X'/10X,'THE BASE LINE SHIFTED FROM AN ABSORBANCE OF ',F6.3,'TO
     3 ',F6.3/5X,110('*')/)
C........IF THERE ARE NO SAMPLES, RETURN TO INITIALIZE FOR A NEW SET
      IF(NSAMP)10,10,80
   80 WRITE(3,85)
   85 FORMAT(1H ,50X,'SAMPLES'/T10,'BOTTLE',T32,'MEASURED',
     1T47,'MEASURED',T63,'TRUE'/ T10,'NUMBER',T30,'TRANSMISSION',T46,
     2'ABSORBANCE',T60,'ABSORBANCE',T75,'CONCENTRATION'/)
C........INPUT FOR SAMPLES
C........VARIABLE......................INPUT COLUMNS....FIELD TYPE....EXAMPLE..
C...
C...BOT   (SAMPLE IDENTIFICATION)            3-6            A4          A15C
C...TRANS  (PERCENT TRANSMISSION * 100.)     7-10           F4.3        0876
C...COM    (COMMENT FIELD)                   11-38          7A4      **REMARKS*
C...
      DO 90 I=1,NSAMP
      READ(2,86)BOT,TRANS,COM
   86 FORMAT(2X,A4,F4.3,7A4)
C........COMPUTE MEASURED AND TRUE ABSORBANCE, AND CONCENTRATION FOR SAMPLES
      ABSM=CONST* ALOG(1./(1.-(1.-TRANS)/RANGE))
      ABS=ABSM-ABASB-((I+NSTD)*DBASE/NUM)
      CONC=(ABS-INCPT)/SLOPE
C........OUTPUT FOR SAMPLES
   90 WRITE(3,95) BOT,TRANS,ABSM,ABS,CONC,COM




   95 FORMAT(1H ,T12,A4,T33,F6.4,T47,F7.4,T61,F7.4,T78,
     1F8.3,T89,7A4)
      GO TO 10
  100 CALL EXIT
      END
```

Instruction for the data input is contained within the program in the form of comments.

Automation of Manual Procedures

Although most methods for automated analysis are based on manual procedures, important changes usually must be made before methods can be successfully automated. When the analyst compares a manual and the corresponding automated method, he will usually find that the ratio of sample to reagents is much greater for the manual than the automated procedure. As has been noted it is desirable to maintain a high flow in the system to obtain optimum response. It is difficult hydrodynamically to join streams of high and low flows, and with small diameter pump tubes frequently it will be difficult to begin liquid flow when starting the pump. Reagents used in automated methods are often more dilute than for the corresponding manual procedure and in some cases the composition must be changed to maintain the correct pH.

The time and temperature conditions specified in a manual method may be changed in an automated method. In many manual procedures fifteen minutes often passes between the addition of one reagent and the next while in automated methods this is rarely the case. In manual procedures, especially when the chemist must analyze many samples, it is difficult, if not impossible, to reproduce exactly the time at which various reagents are added. For this reason, reactions are allowed to go to completion in most manual methods. The excellent reproducibility of the AutoAnalyzer eliminates the need for allowing reactions to achieve completion. Some extremely slow reactions, which take hours of reaction time in manual procedures, can be run at higher temperatures in automated procedures. For methods determining classes of compounds or chemical properties of systems (*e.g.*, total phosphorus, phenols, or COD), different compounds will react with different velocities. When such a method is used, results more precise but different from manual methods may occur. The analyst should make sure that comparable results are obtained for both procedures.

If samples contain higher concentrations than can be determined by a given method, the analyst must dilute the sample. If only infrequent samples contain high concentrations, manual dilution should be used. If all samples to be analyzed are too concentrated, the sensitivity of the method should be decreased by using a smaller diameter pump tube for the sample and including another tube to

pump a compensating volume of distilled water. Another
dilution procedure is possible which makes it convenient
to have a single manifold work for two concentration
ranges. The sample is diluted, segmented with air and
the stream is mixed. This stream is debubbled and a portion of this diluted sample is drawn through another pump
tube to be the sample analyzed in the procedure. To allow
more sensitivity the manifold can be quickly changed by
disconnecting the dilution line. For use without dilution,
water should not be run through the dilution line and the
sampler should be connected to the sample pump tube rather
than to the pump tube used for dilution.

The analysis of chemical oxygen demand (COD) illustrates the use of the AutoAnalyzer modules and the adapt-
ability of the system to samples of different compositions.
In the COD procedure the sample is heated with dichromate,
causing oxidation of the organic matter and reduction of
the dichromate by the reaction:

$$Cr_2O_7^{-2} + 14H^+ + 6e^- \rightarrow 2Cr^{+3} + 7H_2O. \qquad (31)$$

In the manual procedure the excess dichromate is determined by titration with ferrous iron. Since both species
of chromium are colored, the extent of the reaction may
also be measured by the decrease in absorbance due to the
orange $Cr_2O_7^{-2}$ or the increase in absorbance due to the
green Cr^{+3}. The absorbance maximas of $Cr_2O_7^{-2}$ and Cr^{+3}
are 420 and 600 mμ respectively. Measurement of the trivalent chromium at 600 mμ is preferred because in a color
development method--that is one for which the absorbance
increases with increasing concentration of the desired
constituent--the effect of the reagent blank can be evaluated. If, however, great sensitivity is needed, the
decrease in absorbance at 420 mμ should be used.[2]

The flow diagram for the automated analysis of COD as
developed by Adelman is shown in Figure 32. Sample and
dilution water are segmented by air bubbles and mixed.
The diluted sample stream is debubbled and a portion of
this stream is resampled. The dilution obtainable by this
procedure can be as high as ten-fold. The size of the
pump tubes used in the dilution step are chosen to reduce
the COD of the sample into the proper range for analysis
with the only restriction being that the total liquid flow
must be sufficiently large so that no air bubbles are
pulled into the resample line. For analysis of more dilute
samples, the dilution may be omitted and the resample line
connected directly to the sampler probe.

Mercuric sulfate is added to the sample to complex
chlorides since chloride is oxidized by dichromate. The

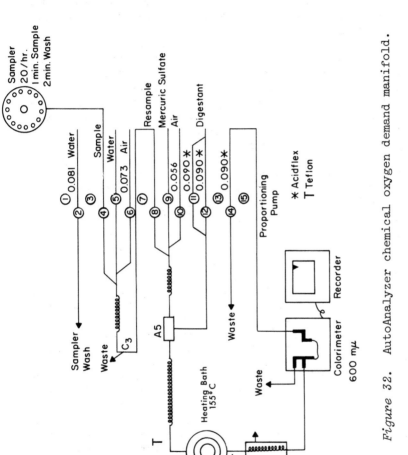

Figure 32. AutoAnalyzer chemical oxygen demand manifold.

total flow of sample plus the mercuric sulfate solution
must equal 2 ml/min. In addition the solubility of mercuric sulfate in the presence of the acid digestant is
limited to 10 mg mercuric sulfate per minute. The ratio
of sample flow to mercuric sulfate flow may be increased,
thus increasing the sensitivity. However, due to the
fixed maximum concentration of mercuric ion tolerable in
the system, the maximum chloride level which may be present in the sample is reduced. The relationship of sample
and mercuric sulfate flow rates, mercuric sulfate reagent
concentration and maximum chloride concentration are given
in Table 23.

The digestant of potassium dichromate in concentrated
sulfuric acid contains a silver sulfate catalyst and is
pumped by two 0.090-inch acidflex tubes. The digestant is
added to the sample stream through a water-jacketed glass
connector. After digestion in a 155°C heating bath, the
stream is cooled in a water-jacketed mixing coil before
being passed through the colorimeter. No pumping of
aqueous solutions should begin until the digestant has
reached the heating bath. Otherwise stream under pressure
will be formed, causing connections to come apart.

Using the ten-fold sample dilution and the smallest
(0.42ml/min) line for resampling, the upper measurable
range of COD is 50,000 mg/l and 23,800 mg/l of chloride
(approximately 5,000 mg/l higher than sea water) may be
tolerated. If the COD is less than 5,000 mg/l and the
chloride is less than 2,400 mg/l, the same system can be
used by disconnecting the dilution step. Switching to 4x
range expansion permits analyses in the range 0 - 500 mg
COD/l. Using a sample flow rate of 1.60 ml/min and a
mercuric sulfate flow of 0.42 ml/min, the range 0 - 160 mg
COD/l can be analyzed using 4x range expansion.

In the development of new analytical procedures and in
the automation of existing methods, the analyst must determine the optimum concentrations of reagents. In manual
procedures the analyst runs samples with several concentrations of the reagent being tested to determine the
range of concentrations giving the same response. Using
the proportioning pump of the AutoAnalyzer, such studies
may be made using a continuous concentration gradient of
the reagent.[62]

In such a study a concentrated solution of the reagent
is pumped into a beaker containing a known volume of
water. The solution in the beaker is mixed by a magnetic
stirrer and the reagent is pumped from the beaker to the
manifold at the same rate as the concentrated reagent
pumped into the beaker. The concentration of the reagent

148

Table 23

Analytical Parameters for the Automated Analysis of Chemical Oxygen Demand[1]

| Sample | | Mercuric Sulfate | | | Maximum Chloride |
Tubing Diameter[2]	Flow (mL/min)	Tubing Diameter[3]	Flow (mL/min)	Concentration (mg/mL)	(mg/l)
0.035	0.42	0.081	1.71	6.25	2,380
0.045	0.80	0.065	1.19	8.33	1,250
0.051	1.00	0.060	1.06	10.00	1,000
0.056	1.20	0.051	0.78	12.50	833
0.065	1.60	0.035	0.43	23.80	625

[1]From Reference 1.

[2]Standard tubing.

[3]Acidflex tubing.

in the beaker at time t is given by the equation

$$C_t = C_c(1 - e^{-\frac{rt}{v}}) \qquad (32)$$

where C_t is the concentration at time t; t is the time, in minutes; C_c is the concentration of the concentrated reagent being pumped into the beaker; r is the pumping rate, in ml/minute; and v is the volume of liquid in the beaker, in ml. Since the pump tubes for input and output used in the system may not be precisely matched, the test should be run over a short time.

Response Characteristics of Flow Systems

Typical recordings for the analysis of chloride are show in Figure 33. The delay time (t_d) between the beginning of sampling and a response on the recorder is independent of the sampling rate. This is the time required for reagent addition and mixing prior to the sample

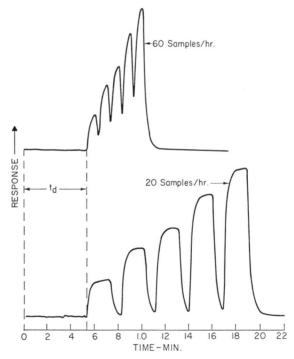

Figure 33. Standard series run at 20 and 60 samples/hour. The delay (t_d) is independent of sampling rate.

150

entering the flow cell. The delay time is dependent on the length of the coils and other components in the system and on the flow rate of fluids. The delay for many systems is 5 - 20 minutes. The total time required for a series of analyses is the delay time plus the product of the number of samples and the sampling rate.

Movement of the sample probe from the wash receptacle to a sample causes a step change in concentration which is maintained to the flow cell because of the prevention of diffusion by air bubbles. The change in recorder response when a sample enters the flow cell is not a step change (Figure 34). The transition time (t_t) or time for the system to reach a new steady-state absorbance is due primarily to mixing in the flow cell.[9] This time will be greater with long flow cells than with short ones. The response time, of course, is also governed by the rate at which the sample is pulled through the colorimeter cell. The response curves for two different diameter pull through tubes are shown in Figure 34. To obtain optimal response, this tube should provide a high flow rate. The maximum size of the pull through tube is governed by the flow rate of the liquid entering the debubbler. To insure that air bubbles do not enter the flow cell, a portion of the liquid must be passed into the waste line at this

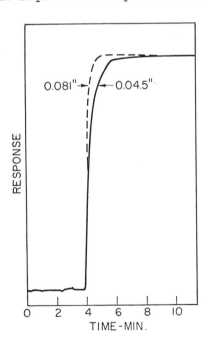

Figure 34. Response of the AutoAnalyzer to a step change in concentration using 0.081- and 0.045-inch tubing to pull the sample through the flowcell.

point, and the maximum flow in the flow cell pull through
tube must therefore be somewhat less than the total liquid
flow. If an analysis has a 0.051" sample tube, 0.040" air
tube, and reagent tubes 0.035, 0.040, and 0.051", the
total fluid flow is 3.62 ml/min, of which 3.02 ml/min is
liquid and 0.60 ml/min is air. The maximum size of the
flow cell pull through tube should be 0.081" (2.50 ml/min)
rather than 0.90" (2.90 ml/min) since the latter might
permit occasional air bubbles to enter the flow cell
causing erratic readings. The effect of a small air
bubble in the flow cell is shown in Figure 35. The transient response has been shown to follow first order
kinetics.[67]

The maximum sampling rate is governed by the length of
the flow cell and the rate at which liquid is pulled
through it. Figure 36 shows the response of a system with
a 50-mm flow cell to three samples using different sampling rates. Samples A and C are the same concentration
while sample B is approximately six times the concentration. At the slower sampling rate the response to samples
A and C is identical but at the higher rate the response
to sample C is significantly higher than to sample A.

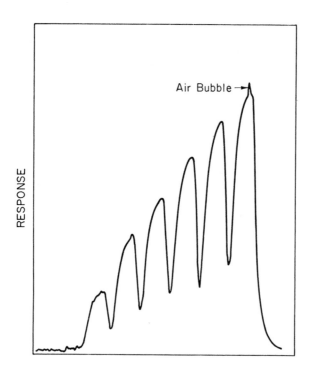

Figure 35.
Effect of an
air bubble in
the flowcell
on the re-
corder
response.

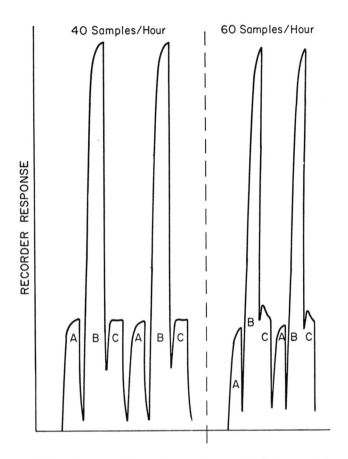

Figure 36. Interaction of samples at high sampling rates.

This is due to insufficient time for flushing sample B from the cell by the wash solution and sample C. The following interaction between samples B and C--

$$\text{interaction} = \frac{c-a}{b} \qquad (33)$$

where a, b, and c are the apparent concentrations of samples A, B, and C--has been shown to be independent of the concentration of the samples used.[68] The analyst may therefore correct for the influence of one sample upon the next. When using 50-mm flow cells, the response will usually approach the steady-state value, but with shorter cells one may use high sampling rates and not achieve a steady-state response. This is possible with the Auto-Analyzer due to the high degree of reproducibility in the processing of samples.

The recorder chart is valuable to the analyst for checking the performance of the instrument and evaluating the precision obtainable in an analytical procedure. A simple test which should be run for each procedure is recording the steady-state response. The sample probe should be placed in a beaker containing a midrange standard or a typical sample and this solution should be pumped through the system for at least ten minutes. Inspection of the chart will indicate the noise level from both instrumental and chemical factors.

Analytical Methods

This section is devoted to the automated analysis of specific substances in natural water. Only one procedure will be shown for many substances although numerous slightly modified procedures exist in the literature. Only colorimetric procedures will be discussed at length, but AutoAnalyzers have been adapted for use with other analytical methods. Fluorometers and flame photometers are available from Technicon. Flame emission and atomic absorption photometers can easily be adapted to the Auto-Analyzer by use of a debubbler, as described by Klein, *et al.*[48] The AutoAnalyzer may simply function to automatically supply samples or it may also dilute the sample or add materials such as lanthanum to suppress the interference of other ions present in the sample. The AutoAnalyzer has been adapted for automated polarographic analysis of discrete samples by Lento.[50] The system removes oxygen from the sample and adds supporting electrolyte and maximum suppressor. The polarograph is operated at a constant potential on the limiting current plateau and the diffusion current is measured as the sample flows through the cell. The automatic electrochemical determination of calcium and magnesium has been described by Fleet, Win and West.[24] By use of a series of valves, EDTA (ethylenediamine tetraacetic acid) or EGTA (ethylene glycol bis-β-aminoethyl ether)-N,N'-tetraacetic acid) is added to distilled water or the sample. The decrease in the anodic wave of EDTA is proportional to the sum of calcium and magnesium while the decrease in the anodic wave of EGTA is due only to the calcium.

Ammonia

Most automated methods for the determination of ammonia utilize the alkaline phenol-hypochlorite reaction. This

was first applied to automated analysis by Logsdon.[51]
Britt[14] used the method to monitor ammonia in the heavy
water (D_2O) of a nuclear reactor, and O'Brien and Fiore[59]
used it in the analysis of sewage. It is believed that
the colored product is an indophenol compound formed by
the reaction[21]

$$NH_4^+ + OCl^- \longrightarrow NH_2Cl \xrightarrow{\text{Phenol}} OC_6H_4NCl \xrightarrow{\text{Phenol}} \qquad (34)$$

$$OC_6H_4NC_6H_4OH \xrightarrow{OH^-} OC_6H_4NC_6H_4O^-$$

Flow diagrams for the determination of ammonia have been
reported by Gaddy,[25] Davies and Taylor,[18] and Anderson, *et
al.*[3] Whitehead, *et al*[72] have modified the method to allow
solids to be separated by a "T" fitting and added EDTA to
prevent precipitation at the high pH of the reaction. The
method can be modified by the addition of nitroprusside to
catalyze the reaction and therefore enhance the sensitivity as suggested by Weatherburn.[71] Using a 5-cm flow
cell, a range of 0-2 ppm N can be achieved without range
expansion.

Grasshoff[29] has automated a procedure using the intermediate reaction product of hypobromite and ammonia to
oxidize iodide which is determined as its blue starch complex. Sawyer and Grisley[64] have used o-tolidine to
determine the chloramine formed by the reaction of ammonia
and hypochlorite.

Chloride

Chlorides may be determined colorimetrically by reacting the sample with mercuric thiocyanate which forms
unionized mercuric chloride and thiocyanate ions. The
thiocyanate is reacted with ferric ion producing highly-
colored ferric thiocyanate proportional to the chloride in
the original sample.

$$2Cl^- + Hg(SCN)_2 \longrightarrow HgCl_2 + 2SCN^- \xrightarrow{Fe^{+3}} Fe(SCN)^{+2} \quad (35)$$

O'Brien[57] has used this method with a dialyzer for the
removal of solids for the determination of 0.5 to 1,000 ppm
chloride in sewage. Britt[14] and Gaddy[25] developed chloride methods with a sensitivity of 5 ppb using 10x range
expansion. Davies and Taylor's method[18] for river water
uses a dialyzer for the removal of suspended solids. The
reported range is 10 to 2,000 ppm using a 15-mm flow cell.
Henriksen[39] developed a method for chlorides in the range
0 to 50 ppm and Anderson, *et al*[3] have developed a procedure for 0 to 500 ppm chloride in sanitary and storm
sewer water using a continuous filter to remove solids.
In Kahn's method[42] mercuric ions are added to tie up a

portion of the thiocyanate released by the chloride in the sample. By controling the amount of mercuric ion and using a dilution, the method is able to determine chloride in the range 30 to 30,000 ppm.

Chemical oxygen demand (COD)

Organic matter in natural water may be determined by oxidation with dichromate. Organic compounds are oxidized to carbon dioxide and water while the dichromate (Cr^{+6}) is reduced to Cr^{+3}. Hexavalent chromium is orange while trivalent chromium is green. The COD may be measured by measuring either valence of chromium. The absorbance at 600 mμ due to Cr^{+3} will increase with increasing COD while at 425 mμ the absorbance will decrease since the unreacted Cr^{+6} is being measured. Automated methods based on the measurement of both Cr^{+3} and Cr^{+6} and using the continuous digestor or a high temperature heating bath (155°C) have been proposed.

Adelman[1] has proposed an automated method using a heating bath and measuring the Cr^{+3} produced by the oxidation of the organic matter. By adjusting the ratio of sample to mercuric ion, which is used to suppress chloride interference, the procedure can analyze COD in a range from 0 to 150 ppm (using 4x range expansion and no sample dilution) or from 0 to 42,000 ppm (without range expansion and using a ten-fold sample dilution). A procedure using the continuous digestor and measuring Cr^{+3} has been reported by Wagner.[70] Ickes, et al[40] have reported procedures using both continuous digesters and 155°C heating baths. Adelman[2] has reported a modification to cover the range 0-20 ppm COD.

Iron

Most natural waters contain low concentrations of iron due to the insolubility of ferric ion, which is the predominant form of iron in oxygenated waters. Most methods for the analysis of iron make use of the intensely colored chelates formed with ferrous ion and 1,10-phenanthroline, 4,7-diphenyl-1,10-phenanthroline (bathophenanthroline), 2,4,6-tripyridyl-s-triazine (TPTZ), 2,2'-bipyridine, or syn-phenyl-2-pyridyl ketoxime.[20] Automated procedures (References 14,25,39,35,61) have generally used TPTZ as the ligand. The structure of the complex formed is[20]

$$(36)$$

Less sensitive procedures utilizing the colored complex of ferric ion and thioglycolic (mercaptoacetic) acid[8] and the complex of ferric ion and thiocyanate[7] have been reported.

Nitrite

The conventional procedure of diazotization and coupling by the Greiss-Ilosvay method or one of its modifications is used for the determination of nitrite. With sulfonilamide and N-1-naphthylethylenediamine as reagents, the reactions are

$$NO_2^- + NH_2C_6H_4SO_2NH_2 + 2H^+ \rightarrow N^+\equiv NC_6H_4SO_2NH_2 + 2H_2O \qquad (37)$$

$$N^+\equiv NC_6H_4SO_2NH_2 + C_{10}H_7NHCH_2CH_2NH_2 \rightarrow \underset{NC_{10}H_6NHCH_2CH_2NH_2}{\overset{N\ C_6H_4SO_2NH_2}{\parallel}} {}_{+H^+}$$

A useful review of spectrophotometric methods for the determination of nitrite has been published by Sawicki, *et al.*[63] Automated procedures for nitrite have been reported by Britt,[14] O'Brien and Fiore,[58] Gaddy,[25] Henriksen,[34] and Kamphake, Hannah, and Cohen.[45]

Nitrate

An automated brucine method has been reported by Kahn and Brezenski[43] for 0.05 to 0.6 mg NO_3-N per liter in estuarine waters. Most AutoAnalyzer procedures for the determination of nitrate are based on manual methods using hydrazine and a copper catalyst,[56] amalgamated cadmium filings,[55] or a cadmium-copper couple[74] to reduce nitrate to nitrite, followed by colorimetric analysis of the nitrite. Automated methods using amalgamated cadmium and a cadmium-copper couple for reduction have been reported by Bernhard and Macchi[5] and Armstrong and LaFond[4] respectively. Brewer, Chan, and Riley[12] and Henriksen[39] have used cadmium filings to reduce nitrate to nitrite. Several investigators have automated the hydrazine method (References 14,25,18,34). The hydrazine reduction method of Kamphake *et al*[45] eliminates the acetone used to destroy excess hydrazine and employs a temperature at which equivalent absorbances are obtained for equimolar nitrate and nitrite standards. The true concentration of nitrate nitrogen is the apparent concentration minus the concentration of nitrite nitrogen.

Kjeldahl nitrogen

The development of the continuous digester has permitted the complete automation of the Kjeldahl procedure for the determination of the nitrogen of organic compounds. The digestion process in the continuous digester has recently been reviewed.[23] The high temperature of the thin film of sample on the wall of the helix permits the digestion to be completed in a few minutes rather than the hours required in a Kjeldahl flask, where the sample temperature does not exceed the boiling point of the liquid. In methods for the determination of nitrogen (References 15,53,66,44), the sample, acid, and catalyst to permit the digestion of refractory nitrogen compounds are pumped into the digester. Following digestion, water is added and the sample is pulled by a second pump into a small mixing chamber and then to a second manifold where the ammonia is determined by the alkaline phenol-hypochlorite method. Kammerer, *et al*[44] have found that the use of 0.01M sodium nitroprusside to dilute the sample after digestion results in a five-fold increase in sensitivity.

Manganese

Henriksen published a method for the determination of 0.02-1.0 mg manganese per liter using a 1.5-cm flow cell.[39,37] The reaction of formaldoxime ($H_2C = NOH$) and manganese in alkaline solution produces the reddish-brown soluble product $(CH_2NO)_3Mn$. The interference of iron is eliminated by adding iron and decomposing the iron formaldoxime complex with EDTA and hydroxylamine hydrochloride.

Phosphate

One of the most important analyses in pollution or productivity studies is the analysis of phosphate. Phosphate is present in many forms in water; usually the analyst will be called upon to analyze for inorganic phosphate, total soluble phosphate and total phosphate. These differentiations are made by filtering and acid digestion of the sample. The phosphate is complexed with molybdate at a pH where silicomolybdate is not formed and the phosphomolybdate is reduced to an intensely blue compound whose absorbance is then measured. Commonly used reducing agents include aminonaphthol sulfonic acid, hydrazine, stannous chloride, and ascorbic acid.

Manual and automated methods using ascorbic acid use potassium antimonyl tartrate to catalyze the reduction. These procedures are sensitive to less than 10 µg phosphorus per liter. Procedures have been automated using ascorbic acid as one reagent and sulfuric acid, ammonium molybdate, and potassium antimonyl tartrate as the second,[74],[28] while other procedures use a single mixed reagent which is prepared daily.[12],[17] Grasshoff[29],[27] has automated a digestion procedure for the determination of organic phosphates. The digestion is achieved by pumping the sample through a quartz coil surrounding a high intensity UV lamp. An automated procedure using a persulfate digestion at 125°C and 3.5 atmospheres pressure has been reported.[6] Excess ascorbic acid was added to react with the large amounts of chlorine released from seawater samples. Alcohol has been added to extend the concentration range of the method.[54]

Stannous chloride reduction methods have been reported by Gaddy,[25] Lee,[49] and Gales and Julian.[26] Gales and Julian also describe a manual digestion procedure for total phosphate compatible with the automated analysis. Henrikson[39],[33] reported an automated method based on the extraction of the phosphomolybdate with isobutanol and subsequent reduction with stannous chloride. In a later paper[36] he reported altering the reagent concentrations to eliminate silica interference. The stannous chloride reduction may be eliminated for the analysis of sewage and highly polluted waters.[39],[38] It has been reported[17] that consistant results were not obtained using Henrikson's method.[33]

Two hydrazine reduction methods for inorganic phosphates have been reported,[3],[7] and a procedure for total phosphate using the continuous digester to hydrolyze the sample with persulfate and concentrated sulfuric acid has also been reported.[7]

Davies and Taylor[18] dialyzed the sample into aminonaphthol sulfonic acid, added acidified molybdate and found a useful concentration range of 0.5-50 ppm. Manual digestion has been used before an analysis using aminonaphthol sulfonic acid as a reductant.[10]

Silica

In all methods (References 39,61,4,28,73,60,13) silica is determined by complexation with acidified molybdate to form a silicomolybdate complex which is reduced to an intense heteropolyblue. Oxalic or tartaric acid is added prior to the reduction to destroy any phosphomolybdate. All procedures are sensitive to low ppb concentrations, and dilution of the sample may be necessary for samples of many natural waters.

Other analyses

Many other analyses have been adapted to the AutoAnalyzer including boron,[19,41] cyanide,[69] iodine,[46] magnesium,[24,47] alkalinity,[18] hardness,[18,32] sulfate,[22] permanganate value,[35] and fluoride (References 25,28,11, 16,31). A distillation column for use with AutoAnalyzer manifolds has been used in the automated determination of fluoride and phenol.[16,52] The determination of copper, chromium, nickel, zinc, and cadmium in waste waters is described by Berry.[8] Automated colorimetric methods have been published for chromium[25] and copper.[39] Analyses for many other water constituents have been automated.[65]

REFERENCES

1. Adelman, M.H. *Automation in Analytical Chemistry; Technicon Symposia 1966*, Vol. I (New York: Mediad, 1967) p 552.
2. Adelman, M.H. Paper presented at 18th Pittsburgh Conference on Analytical Chemistry and Applied Spectroscopy, Pittsburgh, 1967.
3. Anderson, J.J., O.E. Bruss, L.A. Hamilton, and M.L. Robins. *Automation in Analytical Chemistry; Technicon Symposia 1967*, Vol. I (New York: Mediad, 1968) p 373.
4. Armstrong, F.A.J., and E.C. LaFond. Limnol. Oceanog. 11, 538 (1966).
5. Bernhard, M., and G. Macchi. *Automation in Analytical Chemistry; Technicon Symposia 1965*, (New York: Mediad, 1966) p 255.
6. Bernhard, M., E. Torti, and G. Rossi. *Automation in Analytical Chemistry; Technicon Symposia 1967*, Vol. II, (New York: Mediad, 1968) p 395.

7. Bernhardt, H., and A. Wilhelms. *Automation in Analytical Chemistry; Technicon Symposia 1967*, Vol. I, (New York: Mediad, 1968) p 385.

8. Berry, A.J. *Automation in Analytical Chemistry; Technicon Symposia 1966*, Vol. I (New York: Mediad, 1967) p 560.

9. Blaedel, W.J., and R.H. Laessig. *Advances in Analytical Chemistry and Instrumentation*, Vol. 5, C.N. Reilley and F.W. McLafferty, Eds. (New York: Interscience Publishers, 1966) p 69.

10. Blaisdell, D.R., B. Compton, and J.H. Nair, III. *Automation in Analytical Chemistry; Technicon Symposia 1967*, Vol. I (New York: Mediad, 1968) p 393.

11. Bonk, S., and L.N. Scharfe. *Automation in Analytical Chemistry; Technicon Symposia 1966*, Vol. I (New York: Mediad, 1967) p 638.

12. Brewer, P.G., K.M. Chan, and J.P. Riley. *Automation in Analytical Chemistry; Technicon Symposia 1965*, (New York: Mediad, 1966) p 308.

13. Brewer, P.G., and J.P. Riley. Anal. Chim. Acta 35, 514 (1966).

14. Britt, R.D., Jr. Anal. Chem. 34, 1728 (1962).

15. Catanzaro, E.W., G.R. Goldgraben, and R.M. Gasko. *Automation in Analytical Chemistry; Technicon Symposia 1965*, (New York: Mediad, 1966) p 241.

16. Chan, K.M., and J.P. Riley. Anal. Chim. Acta 35, 365 (1966).

17. Chan, K.M., and J.P. Riley. Deep-Sea Res. 13, 467 (1966).

18. Davies, A.W., and K. Taylor. *Automation in Analytical Chemistry; Technicon Symposia 1965*, (New York: Mediad, 1966) p 294.

19. Demmitt, T.F. *Automation in Analytical Chemistry; Technicon Symposia 1965*, (New York: Mediad, 1966) p 204.

20. Diehl, H., and G.F. Smith. *The Iron Reagents: Bathophenanthroline, Bathophenanthroline-Disulfonic Acid, 2,4,6-Tripyridyl-S-Triazine, Phenyl-2-Pyridyl Ketoxime*, 2nd ed. rev. (G. Frederick Smith Chemical Company, 1965).

21. Docherty, A.C. Paper presented at Technicon International Symposium, New York, N.Y., 1964.

22. Ferrara, L.W., R.S. Floyd, and R.W. Blanchar. *Automation in Analytical Chemistry; Technicon Symposia 1965*, (New York: Mediad, 1966) p 109.

23. Ferrari, A., E. Catanzaro, and F. Russo-Alesi. Ann. N.Y. Acad., Sci. 130 (2), 602 (1965).

24. Fleet, B., Soe Win, and T.S. West. *Automation in Analytical Chemistry; Technicon Symposia 1967*, Vol. II (New York: Mediad, 1968) p 355.
25. Gaddy, R.H., Jr. *Automation in Analytical Chemistry; Technicon Symposia 1965*, (New York: Mediad, 1966) p 210.
26. Gales, M.E., Jr., and E.C. Julian. *Automation in Analytical Chemistry; Technicon Symposia 1966*, Vol. I (New York: Mediad, 1967) p 557.
27. Grasshoff, K. Zeit. Anal. Chem. **220**, 89 (1966).
28. Grasshoff, K. *Automation in Analytical Chemistry; Technicon Symposia 1965*, (New York: Mediad, 1966) p 304.
29. Grasshoff, K. *Automation in Analytical Chemistry; Technicon Symposia 1966*, Vol. I (New York: Mediad, 1967) p 573.
30. Hadley, F.C. *Automation in Analytical Chemistry; Technicon Symposia 1965*, (New York: Mediad, 1966) p 383.
31. Harwood, J.E. Water Res. **2**, 637 (1968).
32. Henriksen, A. Vattenhygien **21**, 58 (1965).
33. Henriksen, A. Analyst **90**, 29 (1965).
34. Henriksen, A. Analyst **90**, 83 (1965).
35. Henriksen, A. *Automation in Analytical Chemistry; Technicon Symposia 1965*, (New York: Mediad, 1966) p 301.
36. Henriksen, A. Analyst **91**, 290 (1966).
37. Henriksen, A. Analyst **91**, 647 (1966).
38. Henriksen, A. Analyst **91**, 652 (1966).
39. Henriksen, A. *Automation in Analytical Chemistry; Technicon Symposia 1966*, Vol. I (New York: Mediad, 1967) p 568.
40. Ickes, J.H., E.A. Gray, N.S. Zaleiko, and M.H. Adelman. *Automation in Analytical Chemistry; Technicon Symposia 1967*, Vol. I, (New York: Mediad, 1968) p 351.
41. James, H., and G.H. King. *Automation in Analytical Chemistry; Technicon Symposia 1966*, Vol. II (New York: Mediad, 1967) p 123.
42. Kahn, L. *Automation in Analytical Chemistry; Technicon Symposia 1967*, Vol. I (New York: Mediad, 1968) p 369.
43. Kahn, L., and F.T. Brezenski. Environ. Sci. Technol. **1**, 492 (1967).
44. Kammerer, P.A., M.G. Rodel, R.A. Hughes, and G.F. Lee. Environ. Sci. Technol. **1**, 340 (1967).
45. Kamphake, L.J., S.A. Hannah, and J.M. Cohen. Water Res. **1**, 205 (1967).

46. Keller, H.E., D. Doenecke, and W. Leppla. *Automation in Analytical Chemistry; Technicon Symposia 1967*, Vol. II (New York: Mediad, 1968) p 371.
47. Kelley, P.W., and F.D. Fuller. *Automation in Analytical Chemistry; Technicon Symposia 1965*, (New York: Mediad, 1966) p 266.
48. Klein, B., J.H. Kaufman, and S. Morgenstern. *Automation in Analytical Chemistry; Technicon Symposia 1966*, Vol. I (New York: Mediad, 1967) p 10.
49. Lee, G.E. *Chemical Environment in the Aquatic Habitat*, H.L. Golterman and R.S. Clymo, Eds. (Amsterdam: N.V. Noord-Hollandsche Uitgevers Maatschapij, 1967) p 169.
50. Lento, H.G. *Automation in Analytical Chemistry; Technicon Symposia 1966*, Vol. I (New York: Mediad, 1967) p 598.
51. Logsdon, E.E. Ann. N.Y. Acad. Sci. 87 (1), 801, (1960).
52. Marten, J.F., and D.R. Grady. *Automation in Analytical Chemistry; Technicon Symposia 1966*, Vol. I (New York: Mediad, 1967) p 546.
53. McDaniel, W.H., R.N. Hemphill, and W.T. Donaldson. *Automation in Analytical Chemistry; Technicon Symposia 1967*, Vol. I (New York: Mediad, 1968) p 363.
54. Molof, A.H., G.P. Edwards, and R.W. Schneeman. *Automation in Analytical Chemistry; Technicon Symposia 1965*, (New York: Mediad, 1966) p 245.
55. Morris, A.W., and J.P. Riley. Anal. Chim. Acta 29, 272 (1963).
56. Mullin, J.B., and J.P. Riley. Anal. Chim. Acta 12, 464 (1955).
57. O'Brien, J.E. Wastes Eng. 33, 670 (1962).
58. O'Brien, J.E., and J. Fiore. Wastes Eng. 33, 238 (1962).
59. O'Brien, J.E., and J. Fiore. Wastes Eng. 33, 352 (1962).
60. Place, R., and F. Hardy. *Automation in Analytical Chemistry; Technicon Symposia 1966*, Vol. I (New York: Mediad, 1967) p 649.
61. Rudnik, G. *Automation in Analytical Chemistry; Technicon Symposia 1965*, (New York: Mediad, 1966) p 220.
62. Ryland, A.L., W.P. Pickhardt, and C.D. Lewis. Paper presented at Technicon International Symposium, New York, N.Y., 1964.
63. Sawicki, E., T.W. Stanley, J. Pfaff, and A.D. Amico. Talanta 10, 641 (1963).

64. Sawyer, R., and L.M. Grisley. *Automation in Analytical Chemistry; Technicon Symposia 1967,* Vol. I (New York: Mediad, 1968) p 347.
65. Technicon Corporation, Tarrytown, N.Y., 10591.
66. Tenny, A.M. *Automation in Analytical Chemistry; Technicon Symposia 1966,* Vol. I (New York: Mediad, 1967) p 580.
67. Thiers, R.E., R.R. Cole, and W.J. Kirsch. *Automation in Analytical Chemistry; Technicon Symposia 1966,* Vol. I (New York: Mediad, 1967) p 37.
68. Thiers, R.E., and K.M. Oglesby. Clin. Chem. 10, 246 (1964).
69. Vought, J.H. Plating 52 (5), 420 (1965).
70. Wagner, R. *Automation in Analytical Chemistry; Technicon Symposia 1966,* Vol. II (New York: Mediad, 1967) p 133.
71. Weatherburn, M.W. Anal. Chem. 39, 971 (1967).
72. Whitehead, R., G.H. Cooke, and B.T. Chapman. *Automation in Analytical Chemistry; Technicon Symposia 1967,* Vol. II (New York: Mediad, 1968) p 377.
73. Wilson, A.L. Analyst 90, 271 (1965).
74. Wood, E.D., F.A.J. Armstrong, and F.A. Richards. J. Marine Biol. Assoc. U.K. 47, 23 (1967).

Chapter VIII

Remote Optical
Sensing Techniques

Introduction

The success of a water quality management program or
water resource study will depend in part on the effectiveness of the pollution surveillance and data acquisition
program which is instituted. This fact, together with a
growing public concern regarding environmental quality,
has provided an incentive for the development of instrumentation and techniques for pollution detection, monitoring, and data acquisition. As a consequence,
impressive gains have been made in recent years in the
development of analytical techniques and continuous monitoring instrumentation. However, in view of the
dimensions and dynamic nature of the aquatic environment
it is clear that there is a need for new instrumental
approaches to the problem of surveillance and data acquisition to supplement existing techniques. This need is
particularly evident when dealing with large water masses
such as estuaries, the Great Lakes, and the oceans of the
world. As a result, interest is growing in the concept
of remote sensing.

The term remote sensing refers to the acquisition of
data through the use of sensory devices at positions away
from the subject under investigation. In general terms,
this information is derived from an analysis of a record
showing the interactions of electromagnetic energy with
an object (or substance), or some manifestation resulting
from an interaction. The electromagnetic energy source
may be the sun, in which case sensors depending on solar
reflections are referred to as "passive" systems, or the

energy source may be an "active" source such as a laser. The presentation which follows is restricted to the use of a "passive" airborne system and measurements in the ultra-violet, visible, and infrared regions of the electromagnetic spectrum.

The data acquisition needs in the water pollution control field include information regarding (a) materials on the surface of the water, (b) materials suspended in the water, and (c) materials dissolved in the water. Some of this information is at times expressed in broad-term parameters such as suspended solids, total carbon, oils, etc., and also, of course, in very specific chemical terms. Additionally, there is a need for information of a strictly physical nature, such as temperature or pollution dispersion patterns. Clearly the information needs are quite extensive and as a result cannot be fully realized with a single sensor system. The material presented in the sections which follow deals with one sensor system and presents selected examples of problems wherein remote sensing can offer a number of unique advantages.

General Considerations

The amount of information potentially obtainable from
remote measurement of electromagnetic energy is quite
large. As Colwell, et al,[5] have emphasized, only four
interactions are possible when a photon of any specific
energy strikes a substance. The energy may be absorbed,
emitted, scattered, or reflected. Therefore, in principle, information can be obtained regarding an object
provided, of course, that the instrumentation used is
capable of defining spectral properties in sufficient
detail.

A comprehensive discussion of theoretical principles
relating to remote sensing measurements is clearly outside
the scope of this presentation. Therefore, only a brief
discussion regarding scattering, fluorescence, and thermal
infrared is presented.

Scattering

The process of scattering of light and other electromagnetic radiation has been examined from a theoretical
standpoint by numerous investigators over the years. This
theoretical work is of great practical significance since
it permits a characterization of the "scatterer" from a
measurement of scattering properties. This has led to the
development of techniques for the determination of particle size and concentration of suspended solids, the measurement of turbidity, and the extraction of other
information regarding a "scatterer."[16] The measurement of
turbidity and suspended solids by remote sensing techniques can serve as an illustration of the approach being
utilized in remote sensing technology.

Turbidity is defined in Standard Methods[24] as "an expression of the optical property of a sample which causes
light to be *scattered and absorbed* rather than transmitted
in straight lines through the sample." Based on the above
definition it appears that the term turbidity is directly
related to the expressions "attenuation coefficient" or
"extinction coefficient." Both of the latter terms are
normally used to express the combined effects of absorption and scattering. However, in turbidity measurement
the intent is to measure scattering and absorption over
and above that which is caused by pure water. Hence, the
terms are related but not synonymous.

Both pure and polluted waters will absorb and scatter
light. Generally scattering and absorption are caused by
water molecules and by any suspended particulate materials
present. In the case of pure water, absorption is the

167

dominant mechanism of attenuation. Scattering is generally weak, although the relative importance of each factor is wave length dependent.[9,17]

In the case of natural waters and waste water, absorption generally assumes a lesser role and can even be neglected in turbidity measurement provided pure water attenuation is accounted for in the measurement and color is negligible. Absorption will of course be an important factor in all cases where significant color exists.

From the above Standard Methods' definition of turbidity and the statement of cause of turbidity, it is clear that the dominant property being measured in natural water and waste water is the scattering of light by particulate material. Hence, turbidity may be considered largely a function of the concentration, size, shape, and coefficient of reflection of suspended particles.

From the foregoing statements it is clear that a remote sensing procedure must be designed to measure the intensity of back-scattered light from the target solution and referenced to standard solutions. Ideally, the measurement should be performed within the full spectral region of the "standard candle" (yellow-red regions of spectrum) normally used in laboratory measurement. However, in remote sensing measurements it would generally be desirable to operate in spectral bands which would allow for maximum light penetration. In relatively clear waters, this would allow for the determination of average turbidity of a deep column of water. Only in the case of a bouyant pollutant, would a shift in spectral band be desirable so as to restrict the measurement to a surface measurement. In any event, careful consideration must be given to the attenuation-wavelength relationships involved in selecting operating wavelengths. As indicated earlier, turbidity is an optical property, which in natural waters is largely *due to* suspended solids. Hence, the literature dealing with light scattering by suspended particles must serve as a basis for the development of response-parameter relationships for the remote measurement of turbidity. Fortunately the process of light scattering by particles has been investigated by a number of investigators both in terms of the "single-scatter" approximation and the multiple-scatter theory.[7,16,28,34] Scattering processes have been studied for particles which are small compared with wave length (Rayleigh scattering region) and comparable or large compared with wavelength. This last category is normally treated in terms of the Mie scattering theory.

Three general approaches have been used to formulate the scattering of light and other electromagnetic energy.

It is assumed that the scatterer (1) acts as reflector, or (2) as a refractor or diffractor, or (3) that energy is absorbed by the particle and then re-radiated at the same wavelength.

In the development of mathematical relationships for the characterization of particles in the aquatic environment it has generally been assumed that the Rayleigh scattering theory does not apply, except in very clear natural waters, because of the particle size distribution normally encountered. Secondly, the majority of workers in developing mathematical models have assumed that the particle acts as a reflector and that light is simply reflected from the surface of the particle. As pointed out by Williams,[32] this last assumption is questionable for many natural waters. Only if the particle radius exceeds 10λ does the "reflector" assumption apply quite well.

Several examples may be cited of work which is directly related to the remote sensing problem under consideration in this discussion. For example, Silvestro[23] and Piech, Silvestro and Gray[19] modeled the pollutant-water system of a buoyant fibrous effluent discharging into a river, and measured "pollutant" concentrations using narrow-band aerial photography. The radiance of the effluent stream was assumed to be entirely due to the energy which was scattered or reflected. The ratio of the differential scattering cross section descriptor to the extinction coefficient was correlated with photographic film density as obtained using narrow band aerial photographs. The scattering to extinction ratio was in turn correlated with pollutant concentration and used to develop concentration contours. In order to minimize the scattering and extinction effects of water, the authors worked in the 0.6 - 0.7 μm spectral band. In this region the effects of pure water are very small and the spectral reponse increases with increasing amounts of material in the water.

The work of Williams[30],[31] at the Chesapeake Bay Institute is a second example of work which is particularly relevant to the problem of remote measurement of turbidity and suspended solids in natural waters. The author developed a mathematical model which permits the direct calculation of particle size and concentration from optical data obtained with black and white Secchi discs and a submarine filter photometer. In the development of the model, assumptions were made that the particles were homogeneously distributed throughout the medium, the particles were all the same size, and that they were all spherical in shape. The resulting expressions related the optical equivalent particle size and concentration to

169

extinction coefficient, beam attenuation coefficient, and the relative amount of upwelling light. With some modification, the Williams equations are applicable for use in airborne remote sensing.

Finally the work of Polcyn *et al*,[20] which deals with the problem of measurement of water depth by remote sensing techniques, must be cited. The resulting mathematical models used in depth determination appear to be applicable to turbidity and suspended solids measurements in clear waters, provided some modifications are made in processing and operational techniques. Briefly, the authors developed the following quantitative relationship:

$$\alpha_i - \alpha_j = \frac{1}{(\sec \theta + \sec \phi)\,(Z)} \ln \frac{(V_j/\rho_j)(V_s/\rho_s)_i}{(V_i/\rho_i)(V_s/\rho_s)_j} \quad (38)$$

where subscripts i and j refer to wavelengths i and j, α is attenuation coefficient, θ is viewing angle (from vertical), ϕ is solar illumination angle (from vertical), Z is water depth, V_i, V_j are voltages recorded from sensor outputs at wavelengths i and j, V_s is voltage from sun sensor, ρ_s is reflectance from sun sensor, and ρ_i, ρ_j are scene reflectance at wavelengths i and j. Solution of the above equation would provide values for the attenuation coefficient which is directly related to turbidity and suspended solids.

In practice, use of a laser-scanner system would provide information regarding water depth and electromagnetic power back-scattered to the instrument. All other elements necessary for the solution of the equation are readily determinable. For a more comprehensive treatment of the subject the reader is referred to the original report.[20]

The preceding discussion illustrates that the existing body of theoretical and applied work indicates that turbidity and suspended solids can be measured by an analysis of the degree of back-scattered or upwelling light from a body of water. The instrumental techniques which have shown a potential for this purpose include narrow-band photography, multispectral scanners, and airborne lasers. Examples of each of the above instrumental approaches can be found in the work of Silvestro,[23] Clark, *et al*,[4] Wezernak and Polcyn,[29] and Hickman, *et al*.[11]

Considering the degree of accuracy normally required for turbidity measurement in natural waters, it would appear that requirements can best be achieved by the use of a well-collimated beam of light of high intensity.

170

Hence, the work of Hickman, *et al*,[11] appears to be particularly interesting, and may very well indicate the most satisfactory instrumental approach. The authors, using a pulsed neon blue-green laser, conducted a number of basic transmission-scattering experiments in a 20-ft water tank facility. Included were studies of the variation of peak power with turbidity.

From a theoretical standpoint, an airborne laser-scanner system would appear to be the most suitable instrumental system for the measurement of turbidity and suspended solids. Ideally, this should be a "multispectral" laser system which would operate at a number of wavelengths and also have the capability of being optically scanned. An instrument of this nature would be useful not only for the measurement of turbidity but also for the measurement and detection of a number of other substances in water, including materials which possess the property of fluorescence.

Fluorescence

A number of substances of interest to water pollution control and water resource management can be detected and measured on the basis of their fluorescence properties. These include a variety of industrial wastes, among them such familiar substances as chlorophyll, phenol, ligninsulfonates, Rhodamine WT, and oils. In general, fluorometric methods are more sensitive than absorption techniques, are less affected by turbidity, and usually are extremely useful in the analysis of mixtures. In some instances, only one species in a mixture will fluoresce. In cases where several different consituents fluoresce, analysis is often possible by taking advantage of the additional variable of selective excitation.[6]

A comprehensive discussion of fluorescence and a detailed examination of the spectral characteristics of substances which fluoresce, and are at the same time of importance to water pollution control, is beyond the scope of this presentation. Rather the intent is to discuss the feasibility of airborne fluorometric measurements. With this objective in mind, attention is directed toward measurement of Rhodamine WT. From a remote sensing standpoint, the same considerations involved in the measurement of Rhodamine WT apply to the measurement of other fluorescent substances.

The detection and measurement of Rhodamine WT serves as a good illustration because of the widespread use of this substance in the water resource field. Dispersion studies, determination of mixing patterns and time-of-travel measurements are some of the determinations which are routinely performed using dye tracing techniques. In addition to the above, dye tracing is also used for discharge measurement, studies of circulation and stratification in lakes and reservoirs, tagging of herbicide spray, ground water seepage measurement and many other applications. The procedure is normally carried out using a fluorometer in a continuous sampling mode. The procedures are well known, widely used, and well documented (References 3,8,22,33,35).

Rhodamine WT, Rhodamine B, and Pontacyl Pink are the dyes which are most often used in water resource studies. Rhodamine WT appears to be the first choice for most work because the absorptive losses are smaller as compared to the other two dyes. The spectral characteristics of the three dyes are shown in Table 23.

Table 23

Spectral Characteristics of Selected Dyes

Dye	Excitation Peak	Emission Peak
Rhodamine WT	558 mµ	582 mµ
Rhodamine B	554 mµ	578 mµ
Pontacyl Pink	566 mµ	590 mµ

The value of aerial photography in ocean diffusion studies of dye tracer was discussed by Ichiye and Plutchak.[15] The authors demonstrated that dye concentration can be determined from the photographic negative with a photodensitometer. Their work also shows that synoptic views and rapid repetition of measurement provide greater detail regarding diffusion processes. Unfortunately, this technique is limited to visible concentrations of dye.

The use of airborne fluorometric techniques would, on the other hand, overcome the limitations of photography by providing much greater sensitivity of measurement and by allowing for measurement at levels which are not discernible to the eye. Recent developments in airborne remote sensing instrumentation suggest that airborne fluorometric dye tracing studies are practical. Stoertz, *et al*,[27] using an experimental Fraunhofer line discriminator mounted in a helicopter, were able to monitor Rhodamine WT concentrations of 5 ppb in relatively turbid waters. Controlled experiments over water tanks demonstrated that dye concentrations of 1 ppb were detectable. In this technique, the sun is used as the excitation source; hence, the procedure is limited to daylight hours.

Hickman and Moore[12] investigated the use of a neon-pulsed laser operating at 540 mµ to induce the fluorescence of Rhodamine B. Laboratory studies using this laser, together with a detector system consisting of a RCA 4459 photomultiplier tube and a Tektronix 581 oscilloscope, were conducted. The results indicate that, under favorable conditions, dye concentrations as low as 0.1 ppb are detectable from a height of 100 meters. The authors point out, however, that additional basic research must be performed before the instrument can be considered an operational tool.

The feasibility of detection and identification of fluorescent substances in water, using remote sensing techniques, is presently (1970) still in the research phase. As part of the program in all areas of remote sensing technology at the Willow Run Laboratories, simulated remote sensing experiments with fluorescent substances in water were conducted under laboratory conditions. The resulting data were applied to a hypothetical remote sensing situation in order to calculate signal to noise relationships and determine airborne equipment requirements. The conditions imposed for these calculations were the following:

(a) aircraft flying at elev. of 300 m. (1000 ft.) carrying a laser-multispectral scanner,

(b) velocity to height ratio, $\frac{V}{H} = 0.1$,

(c) illuminated area = 1 m^2 (one resolution element), and

(d) vertical direction of observation.

The signal to noise relationships for these conditions were calculated using the following equation:[2]

$$\frac{S}{N} = (2.55 \times 10^4) \frac{\mu P}{A} \left[\frac{R_c \delta \lambda}{N_\gamma^b + \frac{\mu P}{A} + N_d} \right]^{1/2} \tag{39}$$

where S/N is signal to noise ratio of the return signal, µ is external conversion efficiency, P is laser power, A is area illuminated, R_c is radiant sensitivity of the photocathode, $\delta\lambda$ is resolved wavelength interval used, N_λ^b is spectral radiance from background, and N_d is noise equivalent radiance. In the above equation the term $\frac{\mu P}{A}$ is equal to the spectral radiance due to fluorescence.

The optimum area for laser illumination is one scanner resolution element, or approximately 1 m^2. If the laser is "continuous wave" (cw), its beam will be scanned and a continuous strip image will be obtained as in the conventional passive scanning mode. If, however, the laser is pulsed, and it may well have to be in order to obtain the necessary power output, the type of mapping that can be done will depend upon the rate at which the laser can be pulsed. For the configuration used in the calculations, the dwell time is 4.6 microseconds, a thousand times longer than a typical laser pulse width. The scan mirror takes 30 milliseconds for one revolution. Thus one laser pulse every 30 milliseconds, or a pulse rate of 33.3 per second, would give as an output a strip one resolution element wide, about one meter, for vertical observations, along the flight direction of the airplane. This is well within the capabilities of present laser technology. To obtain a map similar to that obtained with a CW laser if the power were available would require a pulse rate of about 220,000 per second. Since, however, only 6.7 milliseconds of the 30 millisecond revolution time of the scan mirror is spent looking downward, it may be possible to pulse at the 220,000 per second rate for 6.7 milliseconds, that is, about 15,000 pulses, cool the laser for 23.3 milliseconds and start the cycle over again.

The signal-to-noise calculations apply to each resolution element individually and thus are unaffected by matters of laser pulse rates, provided that the laser pulse is longer than the rise time of the detector being used. This condition is easily satisfied. The photomultipliers used in the present scanner have rise times of about 1.6 nanoseconds, or about one-fifth the pulse width of a typical high power laser.

Calculations of S/N for a number of substances as a function of laser power were made. Under favorable conditions the minimum concentrations/thicknesses shown in Table 24 appear to be measurable.

Infrared

All substances above absolute zero (0°K) emit electromagnetic energy, with most objects on earth radiating copiously in the infrared portion of the spectrum. Because

Table 24

External Conversion Efficiencies
(For Given Excitation and Emission Wavelengths)

Substance	λ_e(Å)	λ_f(Å)	$\delta\lambda_f$(Å)	Concentration or thickness	μ	Minimum concentration/thickness	μ min
Rhodamine WT	5461	5900	20	25 µg/l	1.6×10^{-4}	1 µg/l	6.4×10^{-6}
Chlorophyll	4358	6800	260	2.8 mg/m^3	6.4×10^{-4}	.028 mg/m^3	6.4×10^{-6}
Phenol	2752	3000	35	50 mg/l	9.1×10^{-4}	1 mg/l	1.2×10^{-4}
SAE–30 Oil	3655	4500	58	1 mm	3.2×10^{-4}	.01 mm	8.2×10^{-5}
Santa–Barbara Oil	4358	5100	90	.1 mm	1.4×10^{-5}	.01 mm	9.5×10^{-6}

infrared emission occurs throughout a 24 hour period, this portion of the electromagnetic spectrum is of interest since it provides a basis for around-the-clock pollution surveillance.

The positions of the peaks of radiant emittance for various objects are temperature dependent and are defined by the well-known Wien relationship: $\lambda_{max} T \sim 3000 \ \mu°K$. From this relationship it is evident that the radiation peak of the sun (surface temperature approx. 6000°K) occurs at 0.5 μm and that the radiation peak for most earthbound objects (approx. 300°K) would occur at 10 μm. Fortunately an "atmospheric window" exists in this region; hence, thermal infrared sensing is normally performed in the 8-14 μm region.

The radiant emittance of perfect radiators or "black bodies" is described by the Planck and Stefan-Boltzmann laws. However, since few objects even approximate a perfect radiator, the factor "ε" (emissivity) is introduced. Black bodies have an emissivity of 1, whereas other objects will have an emissivity of less than 1, which generally varies with wavelength. Water in the 8-14 μm range, has an emissivity which varies between 0.95 to 0.993.

The increased usefulness of infrared sensing is due primarily to the vastly improved detectors which are currently available and the quality of the optical systems used to collect radiation. Perhaps the most important advances in IR technology have been in the development of photodetectors. In operation some electrical property is monitored which depends upon the number of free charge carriers in the sensitive element.

There are four photoelectric phenomena which are used in infrared detectors. These are photoemission, photoconduction, photovoltaic effect, and the photoelectromagnetic effect. Of these the photoconducting and photovoltaic detectors are most frequently used. For a comprehensive analysis of infrared physics, detectors, and infrared technology, the reader is referred to the work of Holter, et al.[13]

The water resource field has barely begun to exploit the possibilities provided by recent advances in infrared technology. Many scientifically interesting phenomena can be studied with infrared. Selected examples are provided in the sections which follow.

Instrumentation

In a new, rapidly expanding, and highly sophisticated

technology such as remote sensing, a great many instrumental improvements, processing innovations, and new product developments can be expected to occur. Actually the entire subject matter is largely in the research phase of development; hence, new instrumental concepts are constantly being proposed and analyzed. Therefore, only three instrumental systems are described briefly in the sections which follow. These are the Multispectral Sensor, Fraunhofer Line Discriminator, and the Barringer Correlation Spectrometer. Of these, the Multispectral Sensor and the Barringer Correlation Spectrometer may be considered as "operational instruments", whereas, the Fraunhofer Line Discriminator is currently (1970) an experimental instrument.

Multispectral sensor system

The University of Michigan has pioneered the development of a passive multichannel scanner system which is capable of sensing reflected and emitted energy simultaneously in a number of narrow spectral intervals in the ultraviolet, visible and infrared region of the solar spectrum. Basically the system consists of a spectrometer for dispersing the radiation spectrally and filtered detector arrays placed at the focal points of a double-ended optical-mechanical scanner. Figure 37 shows schematically the system operation. The scanner uses reflective optics to collect energy over a wide spectral band between ultraviolet and thermal infrared. The entrance slit of the spectrometer is placed at the focal plane of the scanner to insure that space and time simultaneity is achieved for all channels between 0.4 and 1 μm. Filtered detectors for the infrared region and the use of short electronic time delays provide coincident data for the region beyond 1 μm. Additional detectors and filters can be provided for the ultraviolet region. Signals from the detectors are stored on magnetic tape for later image reconstruction and data processing. As an integral feature of the system, reference lamps, an input proportional to the sun energy, and adjustable temperature plates are viewed each revolution of the scanner mirror. Thus for the first time (with an optical-mechanical scanner) reproducible data can be obtained which is referenced to known radiation sources and is capable of accounting for changes in solar illumination. These controlled points in each scan line permit the use of sophisticated electronic pattern recognition techniques with both analog and digital computers.[18]

The generation of a spectrum for each element permits machine recognition of classes of objects whose spectral

177

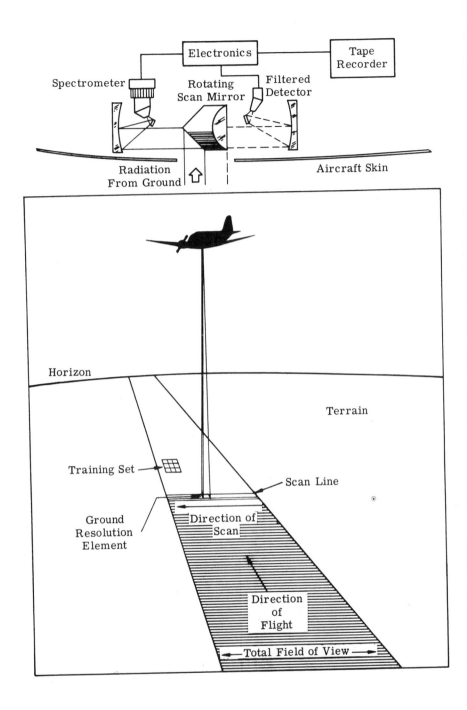

Figure 37. Multispectral Data Collection.

178

characteristics are similar within certain electronic decision levels. As part of on-going research, The University of Michigan is engaged in a spectral measurement program to identify the spectral characteristics of objects and determine their mean and variance statistics for a class of similar objects. In the field of water pollution control, it is believed that remote spectral measurements can be used to detect and identify effluent discharges.

In order to determine spectral characteristics, the airborne data recorded on magnetic tape is electronically digitized and then stored in a computer. The analog to digital converter consists of a high accuracy (13 bit and sign) successive approximation converter capable of providing samples in 10 microseconds. The data is formated and coded into 48 bit words for a CDC-1604 computer. Special computer programs have been written to edit the digital tapes and format the data for use with FORTRAN programs. Selected portions of the objects mapped are used as training areas for which the spectral characteristics are computed. Once computed, a computer-generated spectral curve can be made that is referenced to the incoming solar spectrum or to other objects in the scene. In some cases, ground reflectance standards are used to provide absolute reflectance calibration.

The spectral properties are then available for use with the computer to perform spectral matching operations to enhance the detection of selected objects of similar properties or to map the different concentration of an effluent.

The electronic format of the data also permits a voltage sampling operation (for any selected band) in order to display small density differences. For example, the mixing zones of a thermal discharge into a river or lake can be studied readily by quantizing the data in the 8-13.5 μm band into one-half degree steps, color-coding each step, and combining the set into one color composite display. Each color represents the isotherm for a chosen temperature level. The use of two adjustable reference plates in the scanner systems insures optimum recording and accurate reconstruction of the apparent temperature patterns.

Many other computer programs have been devised and are employed in processing the multispectral data for particular applications.

Fraunhofer line discriminator

The Fraunhofer line discriminator[27] is a relatively recent and potentially very important development in remote sensing instrumentation. This development is significant because it overcomes one of the basic problems associated with outdoor fluorescence measurements by providing a technique for separating relatively weak fluorescence and bright reflected sunlight from the same scene. This is accomplished by working in the Fraunhofer lines of the solar spectrum, wavelengths at which there is little solar energy reaching the earth.

The key elements in the Fraunhofer line discriminator are two narrow band Fabry-Perot filters, which have a half-power bandpass of less than one angstrom, with a peak transmission of greater than 50%. This in effect permits "looking down in" a Fraunhofer line, where reflected solar energy is minimal. Increases in energy within a line will be due to fluorescence of a substance at that particular wave length.

A large number of Fraunhofer lines can be found in the visible, near-U.V., and near-infrared regions of the spectrum. This means that a number of substances can be measured, provided of course that their fluorescence emission peaks coincide with a Fraunhofer line, or at least a significant portion of an emission spectrum overlaps a Fraunhofer line. One example is Rhodamine WT which has an emission spectrum which overlaps the sodium D_2 line at 5890 Å.

In operation, the FLD compares reflected sunlight, which enters at the top of the instrument, with light reflected from the target. This comparison is referenced to any convenient point in the solar spectrum adjacent to the Fraunhofer line. Any increase in energy within a line is related to the intensity of luminescence of the target substance. An analog computer, which is a part of the instrument, compares the ratios found in the solar spectrum (reflected sunlight from above) with the ratios found in the target spectrum.

The present experimental instrument is designed to operate only at one fixed line, the sodium D_2 line at 5890 Å. Undoubtedly in the future this will be expanded to provide for multiband operation.

For a more complete description of the instrument and additional operational details the reader is referred to reports which have appeared recently.[10,26,27]

Barringer correlation spectrometer

The Barringer correlation spectrometer[1] serves as an example of yet another interesting instrumental approach currently being investigated for the measurement of substances in the environment. The method being explored is based on the fact that most vapors and gases have strong absorption bands in the U.V., visible, and infrared regions of the spectrum. Therefore, light passing through a gas or vapor will show absorption bands which are characteristic of the substance being examined. In the Barringer spectrometer, cross-correlation of the incoming spectrum with standards (spectrum correlation masks) serves to identify the gas or vapor being examined and provides information regarding the nature of a substance.

The Barringer instrumental approach appears to be particularly well-suited for air pollution monitoring. In the water field, it provides an interesting approach to the problem of detection of substances and the study of phenomena at the air-water interface.

The fact that many substances have a high enough vapor pressure to liberate significant quantities of vapor in their immediate vicinity indicates that in principle an examination of these vapors should yield information regarding their identity. Of course, this concept is limited to an examination of volatile substances and surface phenomena, *i.e.*, materials floating on the surface of the water, buoyant pollutants, or solutions which are completely mixed.

Preliminary work to date has been directed towards the detection of fish oils, fuel oil, lubricating oils, and iodine vapors. This last category is of interest because of the well-known biological affinities of iodine and its highly volatile nature. Barringer[1] points out that the distribution of surface iodine could serve as an indicator of the distribution of near-surface marine life, a fact of considerable interest to marine biologists.

Selected Examples of Data Acquisition Using the Multispectral Scanner

Detection and identification of any object is possible only if there is a significant contrast of physical or chemical properties between the object being examined and its background. In the case of effluent discharges from power plants, industrial sources, and municipal sources into the aquatic environment, there is normally some significant difference between the effluent and the receiving body of water which can be detected by The University of Michigan multispectral system described earlier. Additionally, as a result of the very large discrimination capability of the sensor system characterization of a target substance is possible, at least in terms of broad parameters. Described in the sections which follow are

181

selected examples which serve to define areas of application wherein a fully developed remote sensing technology could offer a number of advantages.

Temperature Analysis

Temperature is one parameter which is closely related to water pollution and is of interest in virtually all studies dealing with the aquatic environment. Fortunately, surface determinations can readily be made in the 8-13.5 μm region throughout the day or night to an accuracy of 0.5°C. This ability has been clearly demonstrated by a number of investigators. However, the important difference between currently available thermal sensors and the multi-spectral sensor centers around the increased instrument capability due to the calibration and correlation with other portions of the spectrum.

Figure 38 shows a discharge plume from a power plant into Lake Michigan and serves to underscore the dynamic nature and variability of mixing patterns in a large body of water. It also points out the value of large-scale synoptic coverage in many kinds of pollution studies. The area in the middle panel is approximately 1 mile x 1/2 mile.

An added feature of The University of Michigan system is the calibration lamp and controlled thermal sources that are recorded for each scan line. This feature allows for greater sensitivity in temperature determinations and thermal contour development.

Shown in Figure 39 is a portion of a thermal plume discharging from a power plant. The general distribution of the warm water being discharged is clearly visible in the lower panel. Using the computer processing technique referred to earlier, the airborne thermal data was used to develop temperature contours. Shown in the figure are selected temperature levels in the plume area.

Since data collection is in electrical format a number of other processing techniques and forms of presentation are possible. Another example of thermal pollution monitoring and data processing is shown in Figures 40 and 41. In this particular case, data was collected from an altitude of 5000 feet.

In Figures 40 and 41 the lighter areas represent a higher temperature. Shown in Figure 41 is the surface temperature distribution, described in terms of digital symbols which correspond to temperature limits of 0.5°C. The series of numbers along the bottom of the illustration are scan line numbers which also serve as a distance scale. Each scan line represents a distance of 15 feet.

182

7 MAY, 1968, SE WIND AT 14 KTS.

26 SEPTEMBER, 1968, NW WIND AT 10-20 KTS.

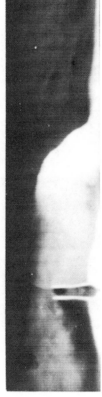

11 AUGUST, 1969, SE WIND AT 4 KTS.

Figure 38. Wind effects on discharge plumes.

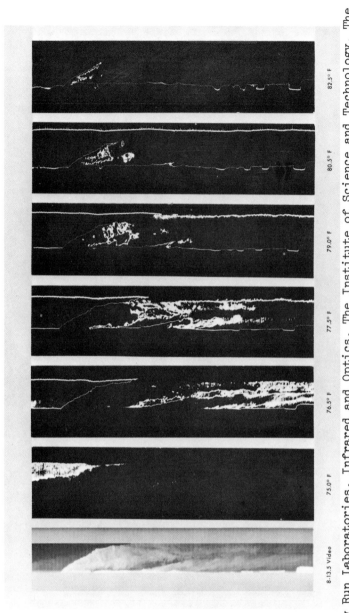

8-13.5 Video   75.0° F   76.5° F   77.5° F   79.0° F   80.5° F   82.5° F

Willow Run Laboratories, Infrared and Optics, The Institute of Science and Technology, The University of Michigan.

Figure 39. Thermal contour of Detroit Edison plant; 13 August 1969; 1130 hrs.; altitude 1000'; sky condition, moderate haze and cloudy.

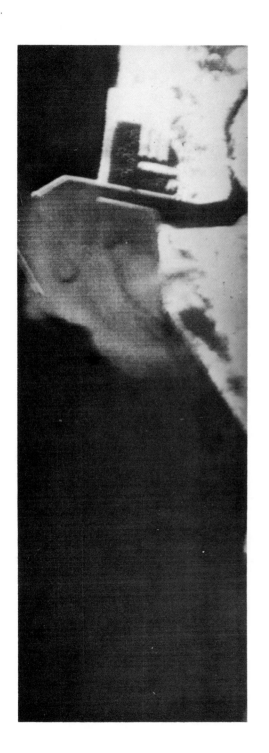

Figure 40. Power plant discharge.

LAKE MICHIGAN WATER POLLUTION STUDY, MICHIGAN CITY, THERMAL

RANGE		SYMBOL	TEMP [°C]
-5.0000	-.1250	▓	▓ = 23.5 C AND BELOW
-.1250	.1250	-	✗ = 23.5 C to 24.0 C
.1250	.3750	=	θ = 24.0 C to 24.5 C
.3750	.6250	*	• = 24.5 C to 25.0 C
.6250	.8750	θ	= = 25.0 C to 25.5 C
.8750	1.1250	✗	- = 25.5 C to 26.0 C
1.1250	5.0000		26.0 C AND ABOVE

Figure 41. Digital thermal contours.

In addition to applicability in thermal pollution studies, the speed, mobility, and ability to completely and virtually instantaneously map large water areas can be of great importance in a variety of other water resource investigations. For example, the movement of water masses can be followed and defined and other phenomena observed. This can be accomplished in terms of a number of physical-chemical characteristics. However, if in a given situation a temperature difference exists between the two water bodies, monitoring can easily be done in the 8-13.5 μm range during hours of daylight or darkness. An example is shown in Figure 42. In this situation the "thermal bar" mixing phenomenon in Lake Michigan is illustrated.[21] The dark area represents water approximately 4°F colder than the light areas shown. This strip of infrared imagery was obtained from an altitude of 12,000 feet and covers a distance along the shore of approximately 30 miles.

Effluent Outfalls

The ability to detect small temperature differences can also be used to locate effluent outfalls including submerged outfalls, provided of course, that the volumes discharged are of sufficient magnitude to alter the temperature of the receiving waters. Detection on the basis of other effluent characteristics is also possible. However, in virtually all cases there is a temperature difference between the effluent and the receiving body of water which can serve as a basis for detection. Monitoring in the 8-13.5 μm region of the electromagnetic spectrum can be done during the day or night and hence provides an expanded capability for pollution surveillance.

Shown in Figure 43 (lower panel) is the plume from a 3 M.G.D. municipal waste treatment plant discharging from a submerged outfall. In this case, the variations in emitted power level are probably due to a change in temperature and emissivity due to a change in water quality.

Oil Pollution

The rapidly expanding need for energy has served as a stimulus for increased oil exploration and oil field development on the land and in the sea. Numerous off-shore wells are currently in operation in various parts of the world, and many more are planned to satisfy growing energy requirements. As a result, the present world oil production stands near 1.8×10^{15} gal./year,[14] and further exploration continues.

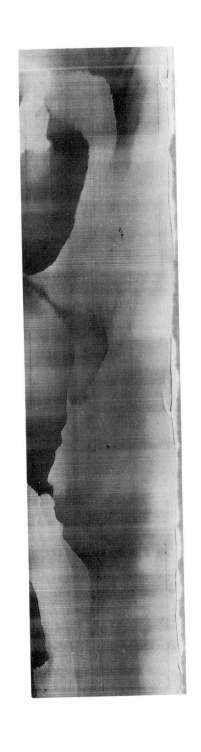

Figure 42. Infrared image of "thermal bar" in Lake Michigan.

Conventional Photo Mosaic

Narrow Band Scanner Imagery 0.55 to 0.58 μ

Enhanced Scanner Imagery 8.0 to 14.0 μ

Willow Run Laboratories, Infrared and Optics, The Institute of Science and Technology, The University of Michigan.

Figure 43. Effluent discharge.

189

Most of the world's oil is transported in huge super-tankers to serve the markets of the world. Considering the very large volumes being transported on the seas, some loss in cargo transfer, accidental spills, leaks, etc. seems inevitable. Even a loss rate as small as 0.1%, which has been estimated as a typical loss in transportation, results in vast amounts being spilled into the sea. In addition there are the accidents, leaks, and spills associated with off-shore oil operations. These factors, together with the fact that some of the new oil field developments are in areas of the world where sea transportation is hazardous, make the problem of oil pollution one of international importance.

The well-publicized Torrey Canyon and Santa Barbara episodes exemplify the magnitude of the problem. Serious damage to water fowl and aquatic life, as well as extensive damage to beaches, occurred from these accidents despite attempts to contain the oil spills. As more and more advances are made in the technology of containment and dispersing of oil slicks, the major concern becomes one of locating oil slicks at sea so that corrective measures can be taken before serious damage is done.

The unfortunate oil leak at Santa Barbara offered an excellent opportunity to observe oil pollution as it would occur naturally. During the period from the morning of 28 January 1969 to midnight of 7 February 1969, oil flowed uncontrolled from a leak in Well 21 of Union Oil's Platform A. According to Holmes,[14] estimates of this initial flow rate disagreed by as much as a factor of 10, but by the 100th day of the spill roughly the same amount of oil that had reached the shores of Cornwall after the Torrey Canyon disaster had been released into the Santa Barbara Channel.

At the request of the U.S. Department of the Interior, personnel of The University of Michigan mapped the oil pollution in the vicinity of Santa Barbara with its airborne multispectral system. The purpose of the flight was (1) to gain some qualitative insight into the problem of locating oil slicks from the air, and (2) to determine the effects and distribution of the dispersant being sprayed on the oil at the time of the surveys.

Data were collected on 7 March 1969 in 17 spectral bands between 0.32 and 13.5 μm. One of the conclusions derived from this study[25] was that no single spectral region in itself provided an optimal remote sensing capability for oil pollution detection and analysis. From the detection standpoint alone, however, the ultraviolet region is the optimum due to the high U.V. reflectance characteristics of oil and the dispersants which are often used.

Figures 44 and 45 illustrate a few of the added advantages of the multispectral approach and data collection in electrical format. Shown in Figure 44 is a two channel display of the scene. The U.V. panel shows the overall distribution of oil and dispersant over the area, whereas the infrared image shows only the thick streamers and areas of oil which were not dispersed. Electronic level slicing of the ultraviolet image provides additional detail by enhancing certain features in a scene. This is illustrated in Figure 45. Panel (a) shows the normal U.V. image of the scene whereas panel (b) shows the contrast between the water and the oil-dispersant plume significantly increased. Panel (c) shows areas of oil, and finally panel (d) clearly defines the oil-dispersant mixture. Therefore, it is evident that by the use of voltage slicing techniques and multichannel examination of a scene a considerable amount of information can be extracted.

Industrial Effluents

As indicated earlier, an important feature of the Michigan multispectral system is the ability to examine a scene in 17 bands distributed between 0.3 and 13.5 µm. This means that in addition to the detection capability referred to in preceeding sections, the system serves as a 17-channel spectrometer with electronic processing features. A few examples of multispectral imagery of industrial discharges are shown in Figures 46-49.

Inspection of the above figures reveals that each effluent is more clearly defined by examination at its optimum wavelength; hence, dispersion patterns of the effluent are best determined within a particular spectral band width.

Using the multispectral scanner concept, the optimum wavelength is automatically determined; in addition, the interactions of radiation with the effluent at other positions in the electromagnetic spectrum are also determined. This is shown in Figures 50 and 51. Each effluent type and receiving body of water has its unique spectral characteristics which serve to differentiate it from other sources.

It is clear that when a photon of energy strikes a liquid, one of the interactions which occurs is scattering. The selective scattering which takes place is, of course, affected by the concentration, shape, and size of suspended solids. A decrease in concentration will result in

Ultraviolet - .32 to .38μ

Infrared - 8.0 to 13.5μ

SANTA BARBARA OIL SLICK

Willow Run Laboratories, Infrared and Optics, The Institute of Science and Technology, The University of Michigan.

Figure 44. Santa Barbara oil slick.

(a) Video (0.32-0.38 μm)

(b) Voltage slice

(c) Voltage slice

(d) Voltage slice

VOLTAGE SLICE OF MAIN OIL SLICK. March 7, 1969, 8:15 a.m. Altitude: 2000 ft.

Figure 45. Voltage slice of main oil slick. March 7, 1969, 8:15 a.m. Altitude: 2000 ft.

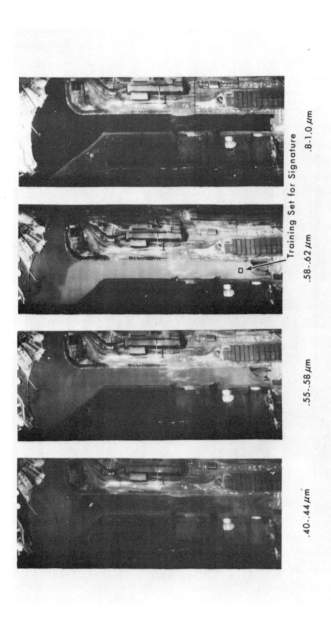

.40–.44 μm   .55–.58 μm   .58–.62 μm   .8–1.0 μm

Training Set for Signature

Willow Run Laboratories, Infrared and Optics, The Institute of Science and Technology, The University of Michigan.

Figure 46. Multispectral imagery of Ford boat slip on Rouge River, Detroit, Michigan; 13 August 1969; 1130 hrs.; altitude 1000'; sky condition, moderate haze and cloudy.

.40-.44 μm   .58-.62 μm   .8-1.0 μm

Training Set for Signature

Willow Run Laboratories, Infrared and Optics, The Institute of Science and Technology, The University of Michigan.

Figure 47. Multispectral imagery of McLouth Steel on Detroit River; 13 August 1969; 1130 hrs.; altitude 1000'; sky condition, moderate haze and cloudy.

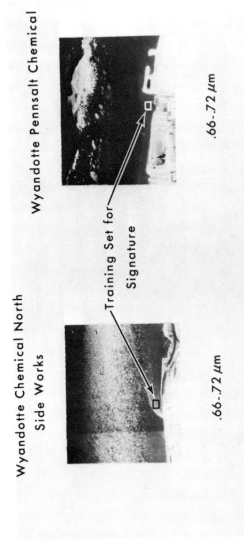

Wyandotte Chemical North   Wyandotte Pennsalt Chemical
Side Works

Training Set for
Signature

.66-.72 μm   .66-.72 μm

Willow Run Laboratories, Infrared and Optics, The Institute of Science and Technology, The
University of Michigan.

Figure 48. Multispectral imagery of Wyandotte North Side works and Pennsalt Chemical; 13 August
1969; 1130 hrs.; altitude 1000'; sky condition, moderate haze and cloudy.

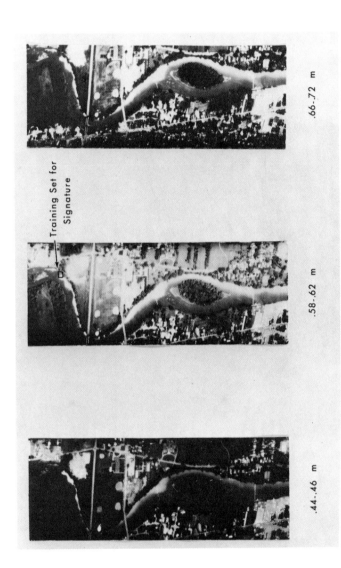

.44-.46 m   .58-.62 m   .66-.72 m

Training Set for
Signature

Willow Run Laboratories, Infrared and Optics, The Institute of Science and Technology, The
University of Michigan.

Figure 49. Multispectral imagery of time container, Raisin River plant; 13 August 1969; 1130
hrs.; altitude 1000'; sky condition, moderate haze and cloudy.

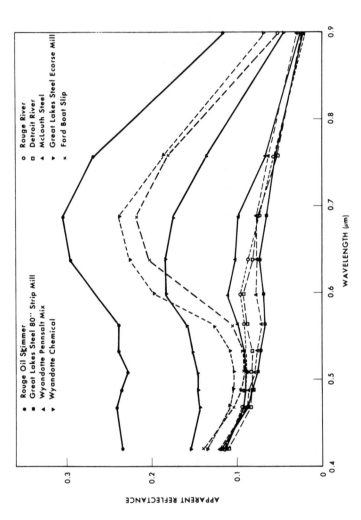

Willow Run Laboratories, Infrared and Optics, The Institute of Science and Technology, The University of Michigan.

Figure 50. Comparison of spectra of major pollutants in Rouge and Detroit Rivers.

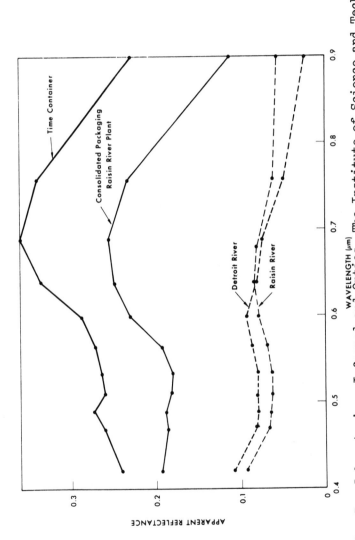

Willow Run Laboratories, Infrared and Optics, The Institute of Science and Technology, The University of Michigan.

Figure 51. Spectra of Time Container and Consolidated Packaging Corporation, Raisin River Plant.

a decrease in back-scattered light. Figure 52 illustrates
this point. The degree of back-scattering is reduced at a
position of 100 ft. below the outfall in the effluent
plume as well as in an adjacent area in the Detroit River.
In this particular case, the optimum spectral position for
defining variations in suspended solids concentration is
about .64 µm. Using the computer techniques referred to
earlier, suspended solids contours could be easily developed. Work is currently in progress to explore further
the technique of suspended solids measurement using the
methods described in addition to other instrumental
concepts.

Conclusions

The added sophistication which the multispectral system
provides in terms of range of spectral discrimination and
electronic processing clearly indicates that this system
can be used to collect a wide variety of data important in
water pollution control and water quality evaluation. As
a result, the preliminary data collected to date encourages
a continued research effort to define fully the capabilities of this system and related instrumentation. Research
is needed to develop spectra for a wide range of pollutants and substances, to define the limits and extent of
ground control required, to define the limitations of the
system, and to evaluate a number of environmental factors.
Additionally, other instrumental concepts must be considered including the use of "active" systems in combination
with the multispectral scanner. This instrumental technique will undoubtedly be further refined and expanded in
the near future. However, it is clear that all measurement needs cannot be fully realized using this approach,
and that additional concepts will have to be explored.
The field of remote sensing is still very much in its
infancy; however, the prospects are very exciting. The
advantages which a fully developed remote sensing technology could offer the field of water quality management are
obvious. Speed of data acquisition, mobility, accessibil-
ity, and large area coverage are some characteristics of
an airborne system which could provide a new dimension in
data acquisition. As a consequence, it is a virtual certainty that in the near future remote sensing techniques
will be used for pollution monitoring, study of dynamic
conditions in the aquatic environment, and other large
scale water resource investigations.

200

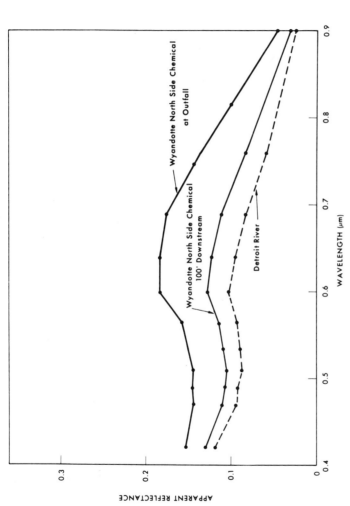

Willow Run Laboratories, Infrared and Optics, The Institute of Science and Technology, The University of Michigan.

Figure 52. Spectra of Wyandotte North Side Chemical showing concentration differences.

REFERENCES

1. Barringer, A.R. "Remote Sensing of Marine Effluvia" in *Oceans from Space*, P.C. Badgley, L. Miloy and L. Childs, Eds. (Houston, Texas: Gulf Publishing Co., 1969).

2. Braithwaite, J. "Dispersive Multispectral Scanning," private communication, Willow Run Laboratories, The University of Michigan, Ann Arbor, Michigan (1970).

3. Butts, T.A. J. San. Eng. Div., ASCE, **95**, (SA 4), 705-714 (1969).

4. Clarke, G.L., G.C. Ewing, and C.J. Lorenzen. *Science* **167**, 1119-1121 (1970).

5. Colwell, R.N., *et al.* Photogrammetric Eng. XXIX, (5), 761-799 (1963).

6. Ellis, D.W. "Luminescence Instrumentation and Experimental Details" in *Fluorescence and Phosphorescence Analysis*, D.M. Hercules, Ed. (New York: Interscience Publishers, 1966).

7. Feinstein, D.L., and K.R. Piech. *Multiple Scatter Effects in Remote Sensing of Environmental Pollution* (Buffalo, N.Y.: Cornell Aeronautical Lab., Inc., 1969).

8. Feverstein, D.L., and R.E. Selleck. J. San. Eng. Divl, ASCE **89** (SA 4), 1-21 (1963).

9. Gazey, B.K. Underwater Sci. Technol. 2, (2), 105-115 (1970).

10. Hemphill, W.R., G.E. Stoertz, and D.A. Markle. "Remote Sensing of Luminescent Materials," in *Proceedings of the Sixth International Symposium on Remote Sensing of Environment*, The University of Michigan, Ann Arbor, Michigan (1969).

11. Hickman, G.D., J.E. Hogg, A.R. Spadaro, and M. Flescher. "The Airborne Pulsed Neon Blue-Green Laser: A New Oceanographic Remote Sensing Device," *Proceedings of the Sixth International Symposium on Remote Sensing of the Environment*, The University of Michigan, Ann Arbor, Michigan (Oct. 1969).

12. Hickman, G.D., and R.B. Moore. "Laser Induced Fluorescence in Rhodamine B and Algae," presented at the 13th Conference on Great Lakes Research, Buffalo, New York (April 1970).

13. Holter, M.R., *et al.* *Fundamentals of Infrared Technology* (New York: The MacMillan Co., 1962).

14. Hoult, D. *Oil on the Sea* (New York: Plenum Press, 1969).

15. Ichiye, T., and N.B. Plutchak. Limnol. Oceanog. II, (3), 364-370 (1966).

16. Kerker, M. *The Scattering of Light and Other Electro-magnetic Radiation* (New York: Academic Press, 1969).
17. Kratohvil, J.P., M. Kerker, and L.E. Oppenheimer. J. Chem. Phys. 43, (3), 914-921 (1965).
18. Nalepka, R.F. *Investigation of Multispectral Discrimination Techniques*, Final Report No. 2264-12-F, Willow Run Laboratories, Institute of Science and Technology, The University of Michigan, Ann Arbor, Michigan (1970).
19. Piech, K.R., F.B. Silvestro, and R.J. Gray. "Industrial Effluent Diffusion in Rivers: A New Approach to Theory and Measurement," Proc. 15th Annual Tech. Meeting of the Inst. of Environ. Science, Anaheim, California (April 1969).
20. Polcyn, F.C., W.L. Brown, and I.J. Sattinger. *The Measurement of Water Depth by Remote Sensing Techniques*, Report 8973-26-F, Willow Run Laboratories, Institute of Science and Technology, The University of Michigan, Ann Arbor, Michigan (1970).
21. Polcyn, F.C., and K. Ewing. "Use of Remote Sensing in Investigations of Circulation Currents in Lake Michigan," Technical Report, Willow Run Laboratories, Institute of Science and Technology, The University of Michigan, Ann Arbor, Michigan (1970).
22. Replogle, J.A., L.E. Myers, and K.J. Brust. J. Hyd. Div., ASCE, 92, (HY 5), 1-15 (1966).
23. Silvestro, F.B. J. Water Poll. Control Fed. 42, (4), 553-561 (1970).
24. *Standard Methods for the Examination of Water and Wastewater*. 12th ed. (New York: A.P.H.A., 1965).
25. Stewart, S., R. Spellicy, and F.C. Polcyn. *Analysis of Multispectral Data of the Santa Barbara Oil Slick*, Final Report No. 3340-4-F, The University of Michigan, Willow Run Laboratories, Ann Arbor, Michigan (1970).
26. Stoertz, G.E., and W.R. Hemphill. "Remote Analysis of Fluorescence by a Fraunhofer Line Discriminator," Proceedings of the Sixth Annual Meeting, Marine Technology Society, Washington, D.C. (1970).
27. Stoertz, G.E., W.R. Hemphill, and D.A. Markle. Marine Technol. Soc. J. 3, (6), 11-26 (1969).
28. Van de Hulst, H.C. *Light Scattering by Small Particles* (New York: John Wiley and Sons, 1957).
29. Wezernak, C.T., and F.C. Polcyn. "Multispectral Remote Sensing Study of Industrial Discharges," *Proceedings of the 25th Purdue Industrial Waste Conference*, Purdue University, Lafayette, Indiana (May 1970).

30. Williams, J. *A Mathematical Model of the Description of the Optical Properties of Turbid Water in Terms of Suspended Particle Size and Concentration,* Technical Report No. 47, Chesapeake Bay Institute, The Johns Hopkins University, Baltimore, Maryland (1968).

31. Williams, J. *Determination of Particle Size and Concentration from Photometer and Secchi Disc Measurements,* Technical Report No. 48, Chesapeake Bay Institute, The Johns Hopkins University, Baltimore, Maryland (1968).

32. Williams, J. *Problems of Ambiguity Involved with the Utilization of the Mie Theory in Particle Size Determination,* Technical Report No. 49, Chesapeake Bay Institute, The Johns Hopkins University, Baltimore, Maryland (1968).

33. Wilson, J.R., Jr. "Fluorometric Procedure for Dye Tracing," *Applications of Hydraulics,* Book 3 (Washington, D.C.: U.S. Gov't Printing Office, 1968).

34. Woodward, D.H. J. Opt. Soc. Amer. 54, (11), 1325-1331 (1964).

35. Wright, R.R., and M.R. Collings. J. Am. Water Works Assoc. 56, (6), 748-754 (1964).

Chapter IX

Physical Characteristics
of Water

Some of the main physical characteristics of natural
and waste waters are (a) density, (b) viscosity, (c) surface tension, (d) temperature, (e) electrical conductance,
(f) turbidity and (g) particulate, volatile and dissolved
solids. These are the major physical parameters of significance in water pollution control. Instrumental measurements of some of the above parameters are discussed
briefly in this chapter.

Density and Viscosity

Density determinations are usually done for the characterization of waste water, *e.g.*, brine waste, for which
concentrations of dissolved solids are commonly indicated
by measuring the density.

Density determinations may be accomplished by measuring
the weight of an exact volume of solution at a given temperature, commonly 20°C (the same temperature at which
volumetric glassware is calibrated). Results are accurate
to 10.0005 g.[2] More commonly, density is measured with a
hydrometer, at the temperature of the sample with appropriate correction to 20°C. Nomographs are usually provided to facilitate conversion.

Density measurements of high salinity waste waters are
also significant for relating concentration on a weight
basis to concentration on a volume basis. By definition,
concentration expressed in terms of parts per million
represents the weight of dissolved matter per one million
equal weights of solution, *i.e.*, milligrams of solute per
kilogram of solution. A concentration of 1,000 ppm will

increase the density of solution by only approximately 0.1 per cent, which may not be significant. At high concentrations, however, corrections must often be made to account for changes in density. The U.S. Geological Survey[2] has arbitrarily selected a concentration level of 7,000 ppm, below which corrections for changes in density are not necessary. Differences between absolute density and specific gravity, and the effect of temperature and salinity on these parameters, have been discussed by Cox.[9]

Viscosity is a direct measure of the resistance of the liquid to flow or fluidity, which is of interest for certain industrial waste waters of high solid content, waste slurries, and sludges. Results are usually expressed in centipoise units at 20°C. The viscosity of pure water is taken to be approximately equal to unity at 20°C (1.009 centipoise).

Several instruments are available for viscosity determinations, *e.g.*, the canal viscometer (Ostwald) or the couette viscometer (or Brookfield viscometer). Viscosity can also be measured by determining the time of fall of a spherical ball of known weight and dimension through a column of the test solution.

Temperature

Temperature is a key parameter which influences all physical, chemical and biological transformations in aquatic environments. It is an intensive quantity of the system under investigation and should be clearly differentiated from heat content which is an extensive quantity.

The effect of changes in temperature on the aquatic environment is presently of great concern to water pollution control agencies. This is due to existing problems of thermal pollution caused by the discharge of cooling water from the power industry into lakes and streams. The discharge of waste heat into surface waters and the resulting increase in temperature will intensify all types of physicochemical and biochemical transformations, the extent of which will depend on the magnitude of temperature change. In addition, the hydrodynamic characteristics of the receiving water may be altered as a result of changes in temperature.

Apart from liquid in glass thermometers, there is a variety of temperature measuring devices. Frequently used in environmental studies is a type of thermometer which may be classified as deformation thermometers, *e.g.*, bimetallic strip.

Temperature measurement by bimetallic strip sensors is based on the physical property of expansion of the two metals. The sensor is composed of two interconnected sheets of different metals which have vastly different thermal expansion coefficients. If this bimetal strip is made in the form of a spiral, it will expand when heated and then can be measured for small temperature changes. The two metals commonly used are brass (copper-zinc alloy) and invar (nickel-steel alloy). Brass has an expansion of over 20 times that of invar. The lag time of this element is about 20 sec or less than half that of the ordinary mercury-in-glass thermometer. Bimetallic strips may be purchased commercially, but they are most commonly used as the sensing elements in thermographs. Commercial bi-metallic thermometers generally cover the range from -40° to +450°C.

Radiation pyrometers are temperature sensing devices based on measurement of radiant energy. In principle, all bodies emit radiant energy at a rate that increases with temperature. A "black body" is a body that absorbs all incident radiation and reflects or transmits none. Hence a black body is an ideal radiator since it only retains at any specified temperature in each part of the spectrum the maximum energy obtainable per unit time from any radiator as a result of temperature alone. Temperature measurement can be made by measuring the emitted radiant energy without making physical contact with the test solution.

Electrical thermometers include such common devices as resistance thermometers, thermistors and thermocouples. Temperature measurement by resistance thermometers are based on the dependency of the electrical resistance of most metals to temperature. Hence, by accurate measure-ment of the resistance, the temperature can be determined.

The higher the temperature coefficient of the resis-tance, the better the metal is for thermoelectric purposes and the higher the sensitivity of the device. Metals used are selected to resist any phase changes during temperature changes. Platinum is frequently used and to a lesser extent, nickel and copper. Resistance measurement can be done by a normal Wheatstone bridge circuitry or by means of more elaborate modifications.[10]

Thermistors are newer transducers commonly used for temperature measurement. These are resistors which are extremely sensitive to temperature. Thermistors are commonly made of sintered mixtures of specially prepared metallic oxides, which when modified in suitable form with platinum alloy wire leads and arranged in a circuit,

207

constitute semiconductors whose electrical resistances vary markedly with temperature.[4]

Thermistors show a high negative coefficient of resistance as a function of temperature. When the logarithm of the resistance is plotted against temperature, the calibration curve obtained is a straight line over small temperature intervals. Thermistors are provided commercially by a number of manufacturers in the form of beads, rods, discs, and probes together with temperature calibration charts. The sensitivity of resistance to temperature varies considerably. However, values of temperature coefficients at ambient temperature, *e.g.*, 25°C, vary from -3.0% to -5.0% per °C. Thermistors time constant in still air varies from 1 to 70 seconds.[12]

Thermistors are frequently used for *in situ* measurement of temperature in natural and waste waters. They also find wide application in conjunction with temperature control and temperature compensation devices.

Another important electrical thermometer frequently used to measure temperature as well as radiation (as a thermopile), humidity and air flow, is the "thermocouple." Basically, this temperature transducer consists of two wires of different metals joined together at both ends. When the two junctions are at different temperatures, an electromotive force is generated, the magnitude of which is dependent on the temperature difference. Hence, if a meter is connected to the circuit, the voltage, and indirectly the temperature difference, is measured.

A detailed discussion on thermocouples can be found in several texts.[6,12] Extensive listing of different types of thermocouples is found in many handbooks. Generally speaking, the reproducibility of thermocouples is not good and they are difficult to calibrate.[6] However, they can be used accurately in the temperature range -50°C to +150°C.

Electrical Conductance

Electrical conductance has been used to give a gross estimate of the ionic strength of the best solution. It has been used extensively in sea water measurement. An excellent review article on this subject has been reported by Cox.[9]

Electrical conductivity

Much confusion exists in the literature regarding interpretation of conductivity data, and the calibration of conductivity salinometers. The electrical conductance, L, of a solution can be represented by the expression

$$L = K_c \sum_i^n C_i \lambda_i Z_i \tag{40}$$

where K_c is a constant, characteristic of the geometry and size of the conductance cell; C is the molar concentration of the individual ions in solution; λ is the equivalent ionic conductance; and Z is the ionic charge. Thus the electrical conductance will vary with the number, size, and charge of the ions, and also with some solvent characteristics such as viscosity. For this reason a meaningful comparison of the electrical conductance of two different types of industrial waste waters may be difficult. Equality in electrical conductance may not mean equality in total dissolved solids. Nonetheless, conductance measurements can be used to good advantage for continuous monitoring of the strength of a given waste water. In this case a change in conductance may be assumed to be due to a change in the number of ions rather than change in the type of ions. Conductance measurements are used extensively for monitoring the quality of surface waters,[14] and in chemical oceanography for salinity determinations.[9]

It should be noted that one of the basic problems in precision conductivity is temperature control. Temperature effects on ionic conductance in heterogenous solutions are quite complex. The temperature coefficient for a solution of constant ionic strength varies with temperature. The conductivity of sea water, for example, increases by 3% per degree increase in temperature at 0°C and only 2% per degree increase at 25°C. At 30°C the conductivity of a solution is about double the value at 0°C. Likewise, the temperature coefficient of conductivity for sea water varies appreciably with large variations in ionic strength.[8,9] In oceanographic work, therefore, relative conductance is determined rather than absolute conductance. In this case the ratio of the conductance of the sample to that of a reference solution at the same temperature is measured. In certain modifications, a thermistor or a resistance is used instead of a reference solution.[9]

Conductivity measurements are quite well suited for *in situ* and continuous type analyses. Great care should be taken, however, to account for changes in temperature, pressure, and other such factors.

Turbidity

The term turbidity is commonly used to signify a visual response to the absorption and scatter of light by suspended matter in a given water. Accordingly, turbidity represents a physiological response to physical optics, similar to color.

Turbidity is measured in terms of the amount of light scattered and/or the light absorbed by matter suspended in the water. The classic procedure for turbidity measurement is based on the preparation of a series of standard suspensions and comparison with the test solution using a Duboscq type comparator.

Measurements are usually done in reference to a standard suspension of fine silica.[1] The standard procedure for the measurement of turbidity[2] is based on the use of "Jackson Candle Turbidimeter."[3,5,11] A schematic diagram of the Jackson Candle Turbidimeter is shown in Figure 53.

Viewer

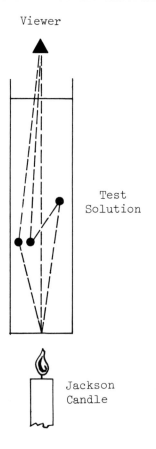

Figure 53. Jackson Candle Turbidimeter.

Test Solution

Jackson Candle

The test is based on measuring that length of light path through the solution at which the outline of the flame of a standard candle becomes indistinct. Results are arbitrarily reported in "Jackson Turbidity Units." This is done in reference to a standard silica suspension, or a formazin polymer suspension which is easier to prepare than the silica suspension.

The turbidity of a given water is a function of both the amount of light absorbed and light scattered by the sample. Light scattered by suspended solids depends on the number, size, shape and refractive index of the particles as well as on the wave length of the exciting light. If the number of particles in suspension is small, particles act as independent scatters and the measured scatter is proportional to the particle concentration. At higher particle concentrations the light scattered is rescattered and interparticle interference occurs.

Instruments used for turbidity measurement by light absorption and light scatter are shown in Figures 54 and 55, respectively. Light absorption measurements are usually applied to water with high turbidity.

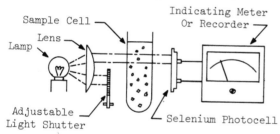

Figure 54. Turbidity measurement by light absorption.

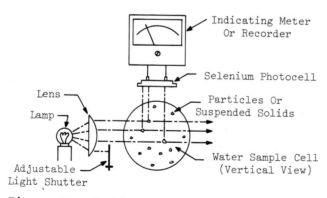

Figure 55. Turbidity measurement by light scattering.

The term "optical turbidity, T" is frequently used[5] as a measure of turbidity in terms of light scattered by a test solution in a given cell light path. This is defined as follows

$$I = I_o e^{-Tb} \qquad (41)$$

or

$$\log (I_o/I) \frac{2.3}{b} = A \frac{2.3}{b} = T \qquad (42)$$

where I_o and I are intensities of incident and transmitted light and b is length of light path. Light scatter measurement are usually made at $90°$ to the incident light beam with a precision in the order of 1%.

Waters of turbidity in excess of 1000 Jackson units are diluted prior to measurement. For waters of low turbidity (less than 25 Jackson turbidity units), light scattering techniques are most commonly used.[1,5] Black and Hannah[5] have discussed the theoretical and procedural aspects of turbidity measurements with the Jackson Candle method and more sophisticated methods. A simple low-angle photometer that may be calibrated with clay suspension in terms of Jackson turbidity units is described and recommended for use with low turbidity waters. Several commercial turbidity monitoring systems are available, and have found wide use for monitoring water quality.[7]

REFERENCES

1. American Society of Testing Material, "Manual on Industrial Water and Industrial Waste Water," *American Society for Testing Materials*, 2nd ed. (1964).
2. APHA, AWWA and WPCF, "Standard Methods - Water and Wastewater," *American Public Health Assoc.*, 12th ed. (1965).
3. Baylis, J.R. Ind. Eng. Chem. 18, 311 (1926).
4. Becker, J.A., C.B. Green and C.L. Pearson. Bell System Technol. J. 26, 170 (1947).
5. Black, A.P., and S.A. Hannah. J. Amer. Water Works Assoc. 57, 901 (1965).
6. Bollinger, L.E. "Proc. Symposium Environmental Measurements - Valid Data and Logic Interpretations," *Public Health Service Publication No. 999-AP-15* (1964) p 41.
7. Cleary, E.J. J. Am. Water Works Assoc. 54, 1347 (1962).
8. Cox, R.A. Deep Sea Res. 9, 504 (1962).

9. Cox, R.A. "The Physical Properties of Sea Water," in *Chemical Oceanography I*, Vol. I, J.P. Riley and G. Skirrows, Eds. (New York: Academic Press, 1965) p 73.
10. Eisner, H.S. J. Sci Inst. 29, 166 (1952).
11. Palin, A.T. Water and Water Engr. 59, 341 (1955).
12. Platt, R.B., and J. Griffiths. "Environmental Measurement and Interpretations" (New York: Rienhold Publishing Co., 1965) p 103.
13. Rainwater, F.H., and L.L. Thatcher. "Methods for Collection and Analysis of Water Samples," in *U.S. Geological Survey Water, Supply Paper 1454* (1960).
14. Wilcox, L.V. J. Am. Water Works Assoc. 42, 775 (1950).

Chapter X

Analysis of Organic Pollutants

Nonspecific Analysis

The chemical oxygen demand (COD) method is widely used as a measure of the organic pollutional load of a water or waste water. It is based on the principle that most organic compounds are oxidized to CO_2 and H_2O by strong oxidizing agents under acid conditions. The measurement represents the amount of oxygen that would be required from a receiving water if oxygen alone could oxidize the organic material to the same end products in the absence of microorganisms. In other words it represents the oxygen that would be needed for aerobic microbial oxidation to CO_2 and H_2O assuming that the organics are biodegradable. The method determines total oxidizable organics but is not capable of distinguishing between those that are biodegradable and those that are not.

Furthermore, the COD test gives no indication of the rate at which organics would be oxidized in a stream. For these reasons, it is dangerous to generally attempt correlation between COD and BOD values, although a relation may be established empirically for a given waste.

The test involves reacting a standard dichromate aliquot in acid solution with a sample containing organic matter until oxidation is complete. Excess dichromate is measured by titration with freshly standardized ferrous ammonium sulfate using Ferroin indicator.

$$14H^+ + Cr_2O_7^= + 6Fe^{++} = 2Cr^{+3} + 6Fe^{+3} + 7H_2O \qquad (43)$$

Any substance that will reduce $Cr_2O_7^=$ will interfere with the COD procedure, leading to high results. The principal interference is chloride ion.

$$Cr_2O_7^= + 6Cl^- = 2Cr^{+3} + 3Cl_2 \qquad (44)$$

This difficulty may be overcome by adding $HgSO_4$ to samples before refluxing to tie up chloride ion as a soluble mercuric chloride complex or by applying a correction of 0.23 x mg/l Cl^- to the analytical result. The latter correction is derived from the stoichiometry of the above interference reaction.

The Total Carbon Analyzer allows a total soluble carbon analysis to be made directly upon an aqueous sample. The result is obtained within a matter of minutes, and the method requires less than 0.5 ml of sample. This volume includes several rinsings of the injection syringe since the actual sample volume required is only 20 µl. If particulate matter is excluded, the test represents total soluble carbon, and if inorganic carbon is excluded or corrected for, the results represent dissolved organic carbon (DOC).

The aqueous sample is injected directly into a combustion tube and heated to 970°C in a constant flow of oxygen gas. Any organic matter is oxidized completely to CO_2 and water vapor on an asbestos packing impregnated with catalyst. These are carried from the combustion tube by the oxygen stream.

Outside the combustion tube, the water is condensed and the CO_2 is swept through a continuous flow sample cell in a nondispersive infrared CO_2 analyzer. The instrument signal is recorded on a linear strip chart recorder, and the peak height or peak area is proportional to the concentration of carbon in the aqueous solution.

A standard curve is established over the concentration range of interest by adjusting the amplifier gain to give midscale response for the midconcentration standard, and by plotting peak height or peak area versus concentration. Commercial instruments are capable of reproducible readings at the 1-2 mg/l carbon level. Peak height for any given carbon concentration is relatively independent of furnace temperature above 940°C, and the same stability is exhibited over a wide range of oxygen flow rates. Standards may be any carbon compound of known purity: Na_2CO_3, potassium acid phthalate (KHP), trishydroxymethylaminomethane (THAM), glucose or benzoic acid. Literature reports[15,18,21,22] are in general agreement that the overall precision is in the order of ±2%.

Inorganic carbon can be removed prior to analysis by acidification and sparging which does not remove volatile acids such as formic and acetic acids.[21] Alternatively, a diffusion cell containing a trough of KOH may be used to trap liberated CO_2 from acidified samples just before withdrawing a liquid sample for carbon analysis.

Biochemical oxygen demand (BOD) is the quantity of oxygen required by a biological community over a period of time to meet its respiratory demands. In this process, reduced organic compounds are oxidized by microorganisms releasing energy for use in synthesis of cellular material which results in population growth. The general equation for aerobic respiration using glucose is

$$C_6H_{12}O_6 + 6O_2 = 6CO_2 + 6H_2O + \text{energy} \qquad (45)$$

For waters with relatively large quantities of organic matter, bacterial metabolism will account for most of the oxygen consumed. But in most natural waters, algae also contribute to the BOD, either by their own cellular respiration or as degradable organic matter, as well as to the rate at which that degradation will occur.

The curve of BOD in organic waste water plotted against time has a steep slope initially but gradually decreases until a plateau is reached as the demand is satisfied. This relation can be described by the equation

$$y = L(1 - 10^{-K_1 t}) \qquad (46)$$

where y is the BOD at time t, L is the ultimate demand when oxidation is complete, K_1 is the rate constant, and t is the time interval after oxidation has begun.

The oxygen depletion curve, then, is a result of organic matter oxidation by bacteria. The resulting growth of the bacterial population is described as logistic or "S" shaped curve with the equation

$$N_t = N_0 \, e^{Kt} \qquad (47)$$

where N_0 is the initial number of cells, N_t is the number of cells after time t, e is the base of natural logarithms, and K is the growth constant.

The rate constant for growth or BOD is affected by environmental factors such as substrate concentration, inorganic nutrient concentration, pH, temperature, toxic compounds and kinds of microorganisms. Thus, although BOD describes the rate of organic matter oxidation under a given set of conditions, the measure of biodegradable organic matter is subject to factors which can result in wide variations in values. To control part of this variability, the BOD measurement is standardized to a considerable extent by holding time, temperature and dilution water constant. But in order to adequately determine the

BOD rate for a given waste or receiving water, reasonably long-term studies are needed in order to quantify the variables in the BOD equation.

Routine measurements of BOD include initial and final dissolved oxygen determination with either an oxygen sensitive electrode or with the Winkler titration procedure, over a 5-day period at 20°C. If the sample contains no microorganisms (industrial wastes), seeding with small volumes of sewage may be required. BOD is then calculated as mg/l of oxygen consumed by the sample under these conditions.

Separation and Identification of Organic Pollutants

This section discusses selected gas-liquid and thin-layer chromatographic procedures used for the detection of specific types of pollutants (pesticides and phenols) as well as compounds of interest in the control of biological sewage treatment processes (digester gas, volatile and fatty acids).

Application of GLC to pesticide analysis

Gas chromatography has been used extensively in the identification and analysis of pesticides in ground, surface and potable waters as well as in sediments, organic tissues, and bottom muds. Although the flame ionization, electron capture and microcoulometric detectors employed for pesticide analysis are quite sensitive, direct aqueous measurement is usually not feasible due to the variety of interfering substances occurring in natural or polluted waters. Furthermore, quantities of sample larger than the detection limits of the detectors are often required for confirming spectral data.

Liquid-liquid extraction procedures are preferable to carbon adsorption for sample pre-concentration.[10] Choices of solvents vary as shown in Table 25 for several organophosphorous pesticides. Chloroform and ether appear to be preferred solvents for the phenoxyalkyl acids, with extractions performed at pH 2.0. Recoveries range from 93-102%.[15,21,22] Neutral pH values generally are optimum for the extraction of chlorinated hydrocarbons.

After extraction, clean-up procedures employing column partition chromatography are necessary. Silica gel columns are most effective for chlorinated hydrocarbons, and florisil columns are used for organo-phosphorous pesticides; elution is performed with a variety of nonpolar solvents.[11] Column, temperature and detector conditions for the separation of three classes of pesticides are summarized

218

Table 25

Solvent Extraction of
Organo-Phosophorous Pesticides[1]

Pesticide	Solvent	No. of Extractions	pH	Mean Recovery %
Parathion	1:1 hexane and benzene	continuous	--	97
Parathion	benzene	1	acid	99-100
Malathion	dichloromethane	3	7.0-8.0	100?
Abate	chloroform	3	1.0	70
Dipterex and DDVP	ethyl acetate	2	8.0	94.5

in Table 26. Additional column data are available in the literature.[11,14]

Retention data are seldom sufficient for the confirmation of pesticide identity. Faust and Suffet[11] have recommended the use of a flow stream splitter prior to flame ionization detection to absorb organo-phosphorous components on KBr. With this procedure it is relatively easy to correlate retention data with confirming spectral data. Infrared spectra may be run on the KBr pellets directly, and ethanol elution of the pellets results in a solution ready for ultraviolet analysis.

Application of GLC to phenol analysis

Adequate separation and resolution of phenols and phenolic acid derivatives are extremely difficult at reasonable operating temperatures due to the activity of these functional groups. Direct aqueous injection is possible and the techniques and column operating conditions for this analysis recently have been reviewed.[4] Acid washing of diatomite supports does not seem to affect phenol analyses by GLC, and salinization of the support leads to increased peak tailing and decreased sensitivity for direct aqueous injection techniques. Carbowax 20M and FFAP (varian aerograph, polyester-type free fatty acid phase) in the 5-10% loading range appear to be the best liquid coatings although many substrates will separate and permit quantification of selected

219

Table 26

Summary of Separation
Conditions for Pesticides with GLC

	Column Packing	Column Length (ft)	Temp. °C	Detector
Chlorinated hydrocarbons	5% DC-200 oil on 60/80 Chromosorb P	6	195	MC
	3% SE 30 on 80/90 Anakrom ABS	4	195	EC
	3% Apiezon L grease on 80/100 Chromosorb W/HMDS	3 or 4	190 or 200	EC
Organo-phosphates	5% DC-200 oil on 60/80 Chromosorb P	6	195	MC
	3% SE-30 on 80/90 Anakrom ABS	4	195	EC
	5% Dow-11 on 60/80 Chromosorb W	6	270	FID
Phenoxyalkyl acids and esters	5% silicone on 80/100 Chromosorb W	6	200	EC
	5% Dow-200 oil on 60/80 Chromosorb W	5	TP; 200-230	MC

monohydric phenols. Some column conditions are given in
Table 27. The combination of FFAP on Chromosorb T provides maximum separation, highest sensitivity and symmetrical peaks with minimum tailing.

Table 27

Column Conditions for
Phenol Analysis

	Column Condition	*Reference*
Phenols from tall oil	Apiezon E, propylene glycol on silica	6
Phenols and cresols	Nylon 66 and 50 HB 2000 on Firebrick	7
Wood phenols	20% Carbowax 20 M on diatomaceous earth	8
t-Butyl phenol mixtures	Silicone oil 550 (3 pts), Carbowax 4000 (2 pts) on Chromosorb W – 220°C	9

Although supplemental identification by spectrophoto-metric or other chromatographic procedures is necessary for complex mixtures, simple phenols may be handled adequately with flame ionization dilution down to 1.0 mg/l without preliminary sample concentration.

Satisfactory results can be obtained by converting phenols and phenolic acids to trimethylsilyl ethers and esters prior to GLC separation to greatly increase volatility, thus allowing operation at much lower temperatures (130-160°C). The dried residue from the solvent extraction step is mixed with a small volume (1.0-5.0ml) of BSA [N,O-bis-(trimethylsilyl)-acetamide]. After allowing time for the completion of the silylation reaction (10-20 min), the liquid produced may be injected directly into a chromatograph. Excellent resolution can be obtained on 3-foot columns of SE 30 silicone gum (10%) on acid-washed chromosorb W at 140°C. Order of elution under these conditions is: phenol < methylphenols < benzoic acid < methoxyl phenol < dihydroxylphenols = methylated acids < methoxylated acids < monohydroxy acids < polyhydroxy acids, in order of increasing retention times.

Application of GLC to volatile acid and digester gas analysis

Gas chromatographic analysis of volatile acids is approximately an order of magnitude more sensitive than older methods involving liquid column partition or paper chromatographic techniques. Most workers[3,13,20] recommend a Carbowax 20M and H_3PO_4 column; although the column compositions vary, 8% Carbowax 20M and 2% H_3PO_4 are probably best. Direct aqueous injection may be employed with a flame ionization detector. The procedure is subject to "ghosting" and often peak areas must be used for quantitative work as peak heights are not linear with concentration. Intermittent water injections may be required to eliminate ghosting, and, although 3-foot columns have been used, longer columns may be desirable. Recent work[21] using Carbowax 20M treated with Tergitol E-68, temperature programming and a carrier gas stream wetted by passing through a $CuSO_4 \cdot 5H_2O$ precolumn has resulted in excellent resolution of all volatile acids (acetic through octanoic) of common interest. With the exception of acetic acid this procedure gives reliable results down to 6.0 mg/l.

Higher fatty acids in sludge samples may be resolved using diethyleneglycol succinate (DEGS), 10% on 60-80 mesh Chromosorb W. A sample is extracted with ether, dried, and the residue converted to the methyl ester by heating in MeOH and 10% HCl for 2.0 hours. After pH adjustment, the esters are re-extracted into ether and injected directly into the chromatograph at 180°C. Lauric, myristic, palmitic and other fatty acids can easily be detected down to 1-5 mg/l.

The volume per cent composition of CH_4, CO_2 and H_2S in sludge digester gas may be determined on columns of silicone grease (28%) on 28-48 mesh C-22 firebrick 70 ft. in length.[19] Thermal conductivity detection is adequate and column temperature is maintained at 30°C. If hydrogen in the gas sample (not separated by silicone column) does not exceed 5 vol. %, the percentage of each gas present is directly related to the relative ratio of sample peak area to total peak areas.

Application of TLC to phenol analysis

Thin-layer chromatography may be the fastest method available for qualitative detection of phenols in water or waste water. As with pesticide analysis, liquid-liquid extraction or freeze concentration should be employed for sample preconcentration. Elution from the plates in ethanol after development is necessary for spectral confirmation.

Plate preparation

A slurry containing 5 gm silica gel to 10 ml water can be prepared in a mortar and poured onto the glass plates that have been lined on all edges with strips of masking tape. Spreading of the layer is accomplished by rolling in one direction with a glass rod (0.5-cm diameter). After drying at room temperature for 15 minutes, the plates can be oven dried at 100°C for 30 minutes, cooled and stored in a dessicator until use.

Uniformity of thickness on any one plate can be obtained remarkably well with a little practice, however, thickness will vary from plate to plate.

Maximum uniformity in thickness of the static phase can be obtained with the use of any of several mechanical applicators that are commercially available.

Application of sample

Most phenols and phenolic acids are soluble in volatile organic solvents such as ethyl ether or 95% ethanol. Generally, the more volatile the solvent, the smaller the spot size after application will be. A gentle stream of dry air directed onto the plate greatly facilitates the application of sample.

Proximity of spots on the same plate is a function of sample concentration, but a minimum horizontal separation of 2.5 cm was found to be adequate in our study. All spots were placed 1-2 cm above the solvent reservoir to prevent draining during development.

Development

(A) Developing chamber

Almost any ordinary vessel can be used: one gallon jars, drinking glasses, or commercially available containers of different sizes covered with plastic caps, glass plates or Dow's Saran Wrap. Solvent can be placed in the vessel which is then capped so that an approach may be made to air saturation with solvent vapor. Plates containing the sample plus *known standard compounds* can then be rapidly inserted for development. R_f values will vary from run to run using this procedure although $R_{standard}$ values are reasonably reproducible.

If reproducible R_f values are desired, constant temperature and air saturation with the solvent system must be provided. A simple means of accomplishing this is to cap the chamber with a two-hole stopper fitted with a stop

cork and a separatory funnel with the tip bent to touch
the inside of the chamber. A prepared plate with the
bottom 3.0 cm free of silica gel can be placed in the
chamber which contains solvent standing to a depth of 1.5
cm. After air saturation is obtained, solvent can be run
in from the separatory funnel to begin development.

(B) Solvent systems

1.	Ethyl ether	---	
2.	Benzene/ethanol	98/2;95/5;90/10	vol/vol
3.	Benzene/methanol/ acetic acid	95/8/4	vol/vol
4.	Benzene/dioxane/ acetic acid	90/25/4	vol/vol
5.	Benzene/acetic acid/water	125/72/3	vol/vol
6.	Isopropanol/ ammonium hydroxide/ water	200/10/20	vol/vol

Detection

(A) Drying

Plates can usually be air dried in 5-10 minutes except
in cases where acid-base indicators are to be sprayed and
the solvent contains an acidic or basic component. In the
latter cases oven drying may be needed.

(B) Fluorescence

Before spraying dried plates should be checked under a
UV hand lamp for any visible fluorescence. *Many* of the
phenolic acids fluoresce brilliantly and their location
can be marked with small indentations in the silica gel
surface. In some cases, exposure to NH_3 fumes greatly
enhances fluorescence.

(C) Chromogenic reagents

Spraying must be done carefully to prevent draining of
the silica gel. Sometimes oven heating enhances color
formation and in other cases the fading or changing of a
color with time may be indicative of certain compounds.
The chromogenic reagents and the structural implications
inferred from them are listed in Table 28.

Ultraviolet and Infrared Identification
of Organic Pollutants

There are relatively few instances reported in the
literature in which UV spectrophotometry has been applied

Table 28

Useful Chromatographic Sprays for the
Identification of Phenolic Compounds

Spray		Structual Implication
Ferric chloride-ferric cyanide spray: 3% $FeCl_3$ in water followed by 3% $K_3Fe(CN)_6$		phenols in general, particularly those of catecholpyrogallol type.
2,4-dinitrophenylhydrazine 0.1% in 2N HCl		aldehydes and ketones.
Diazotized p-nitraniline		
(a) p-nitraniline, 0.3% in 8% HCl	50 ml	phenols in general, particularly mono-
(b) $NaNO_2$, 5% in water	3 ml	hydroxy phenolic
(c) Na_2CO_3, 10% in water	50 ml	compounds.
Tetrazotized benzidine		
(a) Benzidine solution 5 gr. dissolved in 14 ml conc. HCl and diluted to 1 liter	1 volume	phenols in general, particularly those of resorcinolphloroglu-
(b) Aqueous 10% sodium nitrate	1 volume	cinol type.
Vanillin reagent		
(a) Vanillin, 10% in 95% ethanol	2 volume	phloroglucinol-resor- cinol phenols not
(b) Conc. HCl	1 volume	having carbonyl or carboxylic acid groups on the ring or conjugated with it.

directly to natural or polluted waters with effective results. Naturally colored waters show only end absorption
with the cutoff wave length dependent only on path
length.[6] Absorption due to natural color bodies in the
200-350 mµ range is strongly affected by pH, increasing as
pH increases. Some workers[12] have shown that absorption
at 220 mµ is linearly related to COD values. In all such
applications, the nature of the organic compounds was
unknown.

Ultraviolet and infrared spectrophotometric techniques are most often employed to obtain final confirmation of the structure of a component of a polluted sample only after concentration, separation and cleanup procedures have been performed.

Identification of pesticides

Spectrophotometric confirmation of pesticide structure is possible by trapping peaks (split stream) in freeze traps, gas tubes or adsorption on pellets of KBr. In many cases spectra on the eluted fractions in the gas phase from GLC separations are most useful since fine structure is apparent on such spectral traces at low scan speeds. In cases where a component (or fraction) must be handled as a solute in an organic solvent, attention must be given to the absorption characteristics of the solvent. If measurements are made versus air, the solvent cutoff must be known, and if measurements are made versus solvent (usually preferable) wide slit widths may result in poor wave length resolution in regions of solvent absorption.

Qualitative analysis consists of matching the spectra of a suspected component with that of a known pure sample or comparing the wave length positions and intensities of absorption of the component with data listed in several standard reference texts.[5,16,17]

Many of the major IR absorption bands of the organophosphorous pesticides are well documented.[9] A major difference between a parent pesticide and its oxidation product (oxon) is the P=O free band of the oxon at 7.9 μ as shown in Table 29. In general, microgram quantities are required for IR examination, whereas nanogram quantities are sufficient for UV examination (Table 30).

Identification of phenols

Phenolic degradation products of water-borne natural polymers have been characterized by UV spectrophotometry.[7,8] Substituted phenols display sufficiently different UV absorption characteristics to permit identification as shown in Table 31. Bathochromic shifts resulting from base addition to alcoholic solvents are particularly useful for confirming substitution patterns after elution with ethanol from thin-layer plates.

Table 29

Characteristic IR Absorption of
Organo-Phosphorus Pesticides

Group	Wave length (μ)	Intensity*
P = O free	7.9	m
P-O-C aromatic	8.1	m
P-O-C aliphatic	9.7-9.8	s
P-O-ethyl	8.6	w
P-O-methyl	8.4	w
P = S	12-13	m
C-NO$_2$ aromatic	7.4	s
C = O	5.7-5.9	s

*m = medium; s = strong; w = weak.

Table 30

Ultraviolet Absorption Maxima and Limits
of Detection for Selected Pesticides[7]

Compound	λ_{max}	Detection Limit (ng)+
Diazinon	247.5	210
Parathion	274	90
Baytex	252	69
p-nitrophenol	231,313	39
Paraoxon	271	--
Bayoxon	248.5	58

+Determined in 60 μl microcell.

Table 31

UV Absorption Characteristics
of Selected Phenols[7]

Compound	Absorption E+OH	Maxima, mμ E+OH – NaOH
0-dihydroxybenzene	203,219,279	262,290
m-dihydroxybenzene	205,223,277 283	245,295
3-methoxy-4-hydroxy- benzaldehyde	210,255,320	250,350
3-methoxy-4-hydroxy- benzoic acid	259,293	300
3,4-dihydroxybenzoic acid	208,252,292	277,302

REFERENCES

1. Abbott, D.C., *et al.* Analyst 89, 480 (1964).
2. Aly, O.M., and S.D. Faust. Amer. Water Works Assoc. 55, 639 (1963).
3. Andrews, J.F., J.F. Thomas, and E.A. Pearson. Water Sewage Works III, 4, 206 (1964).
4. Baker, R.A., and B.A. Malo. Environ. Sci. Technol. 1, 12, 997 (1967).
5. Bellamy, L.J. *The Infrared Spectra of Complex Molecules*, 2nd ed. (New York: John Wiley and Sons, 1957).
6. Black, A.P., and R.F. Christman. J. Am. Water Works Assoc. 55, 27, New York (1963).
7. Christman, R.F. Environ. Sci. Techol. 1, 4, 303 (1967).
8. Christman, R.F., and M. Ghassemi. J. Am. Water Works Assoc. 58, 6, 723 (1966).
9. Crosby, N.T., and E.Q. Laws. Analyst 89, 319 (1964).
10. Erne, K. Acta Chem. Scand. 17, 1663 (1963).
11. Faust, S.D., and I.H. Suffet. "Analysis for Organic Pesticides in Aquatic Environments," *Microorganic Matter in Water*, ASTM STP 448, American Society for Testing and Materials (1969).

12. Hanya, T., and N. Ogura. Adv. Org. Geochem., Proceedings of the International Meeting, Milan 1962 (New York: Pergaman Press, 1964).
13. Hindin, E., *et al.* Water Sewage Works III, 2, 92 (1964).
14. Lynn, T.R., C.L. Hoffman, and M.M. Austin. *Guide to Stationary Phases for Gas Chromatography* (Hamden, Connecticut: Analabs, Inc., 1968).
15. Minear, R.A. M.S. Thesis, University of Washington (1966).
16. Phillips, J.P. *Spectra-Structure Correlation* (New York: Academic Press, 1964).
17. Rao, C.N.R. *Ultraviolet and Visible Spectroscopy* (London: Butterworths, 1961).
18. Schaffer, R.B., *et al.* J. Water Poll. Control Fed., 37, 1545 (1965).
19. *Standard Methods for the Examination of Water and Waste Water*, 12th ed., APHA, AWWA, WPCF (1965) p 535.
20. Sugar, J.W., and R.A. Conway. J. Water Pollution Control Federation 40, 9, 1622 (1968).
21. VanHall, C.E., D. Barth, and V.A. Stenger. Anal. Chem. 37, 769 (1965).
22. VanHall, C.E., J. Safranko, and V.A. Stenger. Anal. Chem. 35, 315 (1963).

Chapter XI

Analysis for Metal Pollutants

Physicochemical Characteristics of Metal Ions in Aquatic Environments

This discussion is concerned with the nature of the interactions between a metal ion and its aqueous environment, particularly as they may affect analytical procedures. The purpose is not to consider all the sources, transport, controls on and incidences of metal species in the natural water environment; however, some principles of the controlling mechanisms will be considered.

In aqueous solutions of electrolytes there are electrostatic interactions among the charged species which affect the free energy of the ions and, therefore, their chemical activity. In many methods of analysis, such as atomic and molecular absorption spectrophotometry, emission spectroscopy and activation analysis, the detection process is a stoichiometric one in that the signal obtained is directly related to the number of atoms or ions being measured. In other methods, however, particularly electrochemical analyses, the chemical activity, a, is measured. It is related to the concentration, C, through the equation

$$a_i = \gamma_i C_i \tag{48}$$

where γ_i is the activity coefficient, and activity and concentration each have the units of moles per liter.

The principal factor affecting the activity coefficient is the ionic strength, I, also expressible as moles per liter, which is a measure of the total concentration of charged species and is defined as

$$I = 1/2 \ \Sigma C_i \ z_i^2 \qquad (49)$$

where z_i is the charge on ion i. Examples of values of
ionic strength are $3.7 \times 10^{-3}M$ for a typical municipal
water, $7.7 \times 10^{-3}M$ for the Ohio River, and $0.7M$ for sea
water.[1] The effect of ionic strength on the activity coefficient may be expressed through the Davies equation at
25°C

$$- \log \gamma_i = 0.509 \times z_i^2 \ (\frac{\sqrt{I}}{1 + \sqrt{I}} - 0.2 \ I) \qquad (50)$$

Thus the single ion activity coefficient varies both with
ion charge and ionic strength. As the ionic strength
approaches zero, the activity coefficient approaches unity.
Examples of activity coefficients are: for a municipal
water with $I = 3.7 \times 10^{-3}M$, $\gamma_i = 0.94$ for univalent ions
and 0.77 for divalent ions; for sea water with $I = 0.72$ M,
$\gamma_i = 0.7$ for univalent ions and 0.25 for divalent ions.

Thus, in using activity measurements to determine the concentrations of charged species, activity coefficient corrections are frequently necessary.

Another important phenomenon affecting the state of
metal ions is their ability to form hydroxo complexes.
Thus, for example, iron(III) can exist in aqueous solution
as a "bare" Fe^{+++} ion, but also as $Fe(OH)^{++}$, $Fe(OH)_2^+$,
$Fe(OH)_4^-$ and even dimers and polymers such as $Fe_2(OH)_2^{+4}$.[8]
The sole factor affecting the relative concentration of
these monomeric species is pH. Thus at pH 2 or less, the
predominant species is Fe^{+++}, at pH 4 $Fe(OH)^{++}$, at pH
6 $Fe(OH)_2^+$, and at pH 8 and higher $Fe(OH)_4^-$. All of these
various hydroxo complexes of iron are in mutual equilibrium, even though only one species may predominate at a
given pH. One important effect they have is on the solubility of iron(III). Thus as the pH is increased, the
solubility of $Fe(OH)_3$ generally will be much higher than
if such complexes were not formed.

A second important effect of this type of complex formation concerns the optimum condition for an analysis. If
the spectrum of an organic complex of iron(III) is the
basis of the analysis, it is frequently necessary to adjust the pH so as to reduce the competition of OH^- ions
for the iron(III). The same applies when using a chelate
for iron in order to extract it with an organic solvent
prior to analysis. Also when an ion selective electrode
is being used to measure the activity of an ion in solution, such as the divalent ion electrode for Fe(II), it
will detect the Fe^{++} form. If some of the Fe(II) is in

other soluble states, such as hydroxo complexes, that portion will not be detected. Generally, the higher the oxidation state of a metal, the more readily it forms hydroxo and other complexes. These are not usually important species for the alkali metal ions.

A great variety of other metal ion complexes can form in natural and treated waters. Aluminum is readily complexed by fluoride to form such species as AlF^{++} and iron(II) also forms complexes with fluoride. Tannin, which is found in many plants, hydrolyzes in water to form digallic acid (tannic acid), which is an aromatic carboxylic acid, a weak acid with a pH of between 5 and 6. It has been shown that tannic acid forms a relatively strong complex with Fe^{++}.[12]

In measurements of calcium activity in sea water, it was shown that approximately 16% of the calcium is not ionized but is complexed or bound as ion pairs;[27] similarly, it was shown that 10% of the magnesium is bound, probably by sulfate, carbonate and bicarbonate.[26] Ion pair formation can occur in relatively concentrated solutions and can be considered as another kind of metal complex. Thus, for example, the interaction of Ca^{++} and HCO_3^- to form $CaHCO_3^-$ or soluble $CaCO_3$ is essentially ion pair formation. In a typical municipal water, it has been calculated that perhaps only 0.1% of the calcium forms the ion pair $CaCO_3$; similarly, 1.5% may form $CaSO_4$.[1] Even univalent cations can form ion pairs such as $NaSO_4^-$ and KSO_4^-.[10] As in the case of hydroxo complexes of metals, ion pairs and other complexes can affect activity measurements, as well as influence their ability to be detected by other means or be separated or concentrated to facilitate analysis.

Another important state of metals in water is the colloidal form. Thus, many crystalline or amorphous metal hydroxides, carbonates, silicates, and other species, as well as inorganic polymers, can exist in the colloidal state and remain relatively stable for long periods of time. In addition, soluble monomeric metal ions can adsorb and thereby associate with these colloidal species. All of these colloidal forms of metals can be significant sources of their total analyzable concentration. They will not settle out from solution unless coagulated; they may, however, be removed with fine filters in order to distinguish them from truly soluble species. It sometimes requires several hours of digestion in concentrated acid to convert them to a soluble and, hence, analyzable state.

A final important variable affecting the state of a metal ion species is the redox potential of the solution.

Thus, for the reduction reaction

$$Fe^{+3} + 1\ e \leftrightarrow Fe^{++} \tag{51}$$

the equilibrium activity ratio between these two iron species is determined by the reduction (redox) potential of the solution Eh, the "h" referring to the fact that this is the potential compared to the normal hydrogen potential taken as zero. The equilibrium relationship at 25°C is

$$Eh = E^O - 0.059\ \log \frac{(Fe^{++})}{(Fe^{+3})} \tag{52}$$

with E^O as the standard reduction potential for this reaction in volts, and ( ) refers to ionic activities. This Nernst equation indicates that the more positive Eh, the smaller the ration of $(Fe^{++})/(Fe^{+3})$, provided that the system is at equilibrium. Generally, the reduction potential of natural waters increases with oxygen concentration, in which case (Fe^{+3}) should increase as well.

There are two principal difficulties in utilizing this concept and relating it to the behavior in natural waters.[19] First, it is difficult to measure reduction potentials, and second, there is no assurance, once an accurate measurement is made, that the redox system of interest is in equilibrium with it. Nevertheless, the fact that two or more oxidation states can exist for the same metal can have a profound effect on the chemical analysis of that metal. Generally, the chemical behavior of the various oxidation states is quite different, including their solubilities and ability to be complexed and chelated, extracted and precipitated. Also their molecular absorption spectra are different. To simplify the measurement of the total concentration of such a species, it is usually convenient to convert all forms into one oxidation state.

It is apparent from this discussion that the physico-chemical state of a metal ion can have a great effect on almost every analytical process, with the possible exception of the measurement of radioactive species. It is especially important to keep this point in mind when one considers the common instrumental techniques for metal analysis in aqueous systems.

Molecular Absorption Photometry and Colorimetry

The most widely used techniques for the analysis of metals in water, other than the alkali and alkaline earth metals, involve the absorption of light by the metals in the form of soluble complexes, chelates, compounds, or "lakes." Although many of these metals at high

concentrations do absorb light, at the levels to be found
in environmental, municipal, industrial and waste waters
it is often necessary to form a more highly light-adsorb-
ing species (to use a "color developing agent") which
presumably and optimally is specific for the test metal
of interest.

Many of these current techniques for metals involve
species which adsorb visible light. Table 32 lists such
methods for thirteen metals, showing the minimum detectable amounts and some important interferences. The optimum wave length in each case is in the visual region of
the light spectrum, and this permits the use of both
visual colorimetry and instrumental measurement, such as
with filter photometers and spectrophotometers. Table 10
has similar information for these and other metals.

One of the principal advantages of the use of such
color developing agents resulting in light absorption in
the visual region is the fact that visual colorimetry can
be utilized. Also it is not necessary to resort to instrumental detectors. However, usually the use of a spectrophotometer can improve the sensitivity and accuracy of the
method. One of the principal disadvantages of many of
these techniques is that there are interferences which, in
some cases, require a high degree of chemical sophistication and intuition to eliminate. An alternative approach
is to resort to more expensive instrumental detectors,
such as atomic absorption and emission spectroscopy, where
the interferences may be substantially fewer.

Although chemical means or preseparating techniques,
such as ion exchange chromatography, may be used to reduce
interferences, spectrophotometry can also be utilized.
When the interference is due to overlap of absorption
spectra, spectrophotometry can correct for such overlap,
thereby removing the interference. An example of such a
method is the analysis of nickel in industrial water,
using the yellow-green complex of nickel with diethyldithiocarbamate.[6] Copper, cobalt and bismuth also form
complexes with the latter which absorb light at 328 mμ,
and corrections must and can be made by determining their
concentrations separately. If copper is the sole interfering metal, two measurements can be made on the unknown
sample, one at 328 mμ, the absorption maximum for the
nickel complex, and one at 436 mμ, that of copper. The
two measurements are required because each of these respective complexes adsorbs light to some extent at the absorption peak of the other. After measuring at these two wave
lengths, the absorbance of standards containing known mixtures of copper and nickel, the concentration of copper

Table 32

Metal Ions Determined by Colorimetry and Molecular Absorption Spectrophotometry*

Metal	Color Developing Agent	Optimum Wave Length	Minimum Detectable Conc.	Possible Interferences
		mµ	µg/l	
Al	"Aluminon"	525	2	F, Ca, chlorine, sulfite polyphosphate
As	Diethyldithio-carbamate	535	1 µg	Co, Ni, Hg, Ag, Pt, Cu, Cr, Mo, Sb
B	Curcumin	540	0.2 µg	nitrate
Cd	Dithizone	518	0.5 µg	Pb, Zn, Cu
Cu	Cuprethol	435	0.2 (visual)	Bi, Hg, Co, Ni, Ag, etc.
Cr	Diphenylcar-bazide	540	10	V, Fe, Mo, Cu, Hg
Fe	Phenanthro-line	510	3 µg	Cr, Zn, Co, Cu, Ni, etc. oxidants
Pb	Dithizone	510	2 µg	heavy metals, organics
Mn	Persulfate	525	5 µg	Br, I
Ni	Heptoxime	445	----	----
Se	Diaminoben-zidine	420	1 µg	Fe, I, Br, oxidants
Ag	Dithizone	462 620	0.2 µg	Fe, Cl_2, oxidants
Zn	Dithizone	535	----	Cd, Cu, Pb, Ni

*Standard Methods for the Examination of Water and Waste Water, 12th ed., (New York: American Public Health Association, 1965).

and nickel may be calculated separately from the two measurements on the mixture of the test solution by simply solving two simultaneous equations.

Although many of these colorimetric and photometric techniques have been used for many years and may require relatively inexpensive or virtually no instrumentation, in some cases they are as accurate, precise and sensitive as emission spectroscopy, atomic absorption spectrophotometry and neutron activation analysis, which require relatively expensive instrumentation.

Atomic Absorption Spectrophotometry

Atomic absorption spectrophotometry (AAS) has wide application in the analysis of metal ions in natural and treated waters and waste effluents because of its speed, low cost per analysis, simplicity, and frequent ability to analyze complex mixtures without prior separation. A wide variety of applications have been reported, including analyses of industrial water and waste water,[4,20] fresh water,[9] sea water,[5,15] and sewage.[25]

The sensitivity of atomic absorption is generally considered to be that concentration of test species producing one per cent adsorption of light of the appropriate wave length passed through the flame into which the test solution is being aspirated.[16] One definition of *detection limit* is that concentration giving a detection signal twice the variability of the background. In natural waters, many metals frequently can be analyzed directly, because they are present at a high enough concentration to exceed the sensitivity limit. Examples are sodium, potassium, calcium, magnesium, strontium, lithium, manganese, copper and zinc.[9] In other cases, AAS is not sensitive enough, and preconcentration and solvent extraction are required. These would be needed for copper, lead, zinc, nickel, iron and cadmium in industrial waters,[4,20] cobalt, nickel and lead in fresh natural waters,[9] and iron and nickel in brine.[15]

Several preconcentration techniques, including coprecipitation, ion exchange and solvent extraction, have been utilized for the analysis of metals in water. The advantage of the latter is that in addition to concentrating the metal, AAS has a higher inherent sensitivity for many metals in an organic solvent than for those in water. For example, one study has indicated that the sensitivity in methyl isobutylketone (MIBK), a common solvent used for this purpose, compared to water was greater for metals in the range of 0.25 to 1.0 mg/l by the following factors:

Fe 3.8, Cu 4.6, Pb 6.3, Zn 3.0, and Cd 2.6.[20]

The limits of detection for several metals in water with one instrument are shown in Table 33, along with a comparison for the similar limits obtained when the metals are first chelated in water, using a chelating agent such as sodium diethyldithiocarbamate, followed by extraction with MIBK. This method frequently can improve the sensitivity by as much as a factor of 50. In one of the studies of chelation and extraction prior to AAS analysis, it was found that the proper choice of chelating agent before extraction could increase the sensitivity (*e.g.*, oxime for iron and dimethylglyoxime for nickel). This study also showed that the presence of salt in brine reduces the sensitivity by a factor of 2.[15]

Table 33

Comparison of Detection Limits for
Several Metals with Atomic Absorption Spectro-
Photometry, Directly in Water and Using
Solvent Extraction

	Detection Limits (µg/1)			
		Solvent Extraction		
Metal	*Direct in H_2O (16)*	*Ref. 20*	*Ref. 9*	*Ref. 15*
Fe	5	2–3		0.2
Cu	5	0.2		
Zn	2	0.3		
Pb	30	5	0.8	
Cd	1	0.7		
Co	5		0.4	
Ni	5		0.4	1.0

The state of the metal ion in solution can sometimes reduce the ability of AAS to detect it, as can the presence of other species in solution. Such interferences can generally be classified as "chemical," "ionization," and "matrix."[16] A chemical interference can occur when a compound of the test element is formed in the flame, preventing the test species from being reduced. One example is the interference of silicon with a calcium analysis: the

238

addition of lanthanum binds the silicon, freeing the calcium. Ionization interferences result when the element in
the flame is ionized, absorbing a different wave length;
this can be reduced by flame adjustment. The presence of
large amounts of dissolved solids can reduce the light
absorption, a matrix interference essentially due to a
reduction in solution flow through the burner. Thus, high
phosphates of the order of 1.0M can reduce the signal in
copper and magnesium analysis by 20%.[16] Finally, it
should be noted that undissolved particulates, such as
iron oxide, will generally not contribute to light absorption. Thus, to determine total iron in a sample, the
particulate iron should be dissolved before AAS analysis.[20]

Other Flame Techniques

Although atomic absorption spectrophotometry is now
widely used for a great many metals in water analysis,
nevertheless flame photometry is still a useful technique,
especially for the alkali and alkaline earth metals. More
sensitivity, however, can be obtained for the former. For
other metals, flame photometry can also be used, although
the sensitivities are sometimes poorer than for colorimetric and other instrumental techniques.

For the determination of many alkali and alkaline earth
elements by flame photometry in such a multicomponent system as sea water, it is generally necessary to separate
the test ion before analysis because of interferences from
major elements, like sodium, as well as minor elements,
where emission spectra overlap with that of the test ion.
Also for most accurate analysis, it is often necessary to
add a known amount of an internal standard to the test ion
solution to eliminate variability in the flame
characteristics.

In one scheme of analysis for the major cations in sea
water, the test solution was adsorbed onto a cation exchange resin which was eluted by a successive displacement
technique. After being separated from the potassium,
sodium, magnesium, and calcium, the strontium was then
analyzed by flame photometry.[11] Lithium has been analyzed
similarly in sea water after being separated from other
alkali metals and alkaline earths,[24] as have other such
ions as potassium,[21] strontium and barium.[2]

In addition to interferences caused by emission from
other elements, there are also interferences resulting
from suppression of emission and the effects of foreign
constituents on solution viscosity, surface tension and
volatility. Nevertheless, with care and the use of

separation techniques when necessary, flame photometry can be a useful analytical tool, especially for the alkali metals.

Another potentially useful flame technique which has been developed in recent years is atomic fluorescence flame spectrometry.[7] This method depends on the excitation of neutral atoms in a flame by an outside source of light, such as a xenon arc lamp, and the subsequent measurement of the emission intensity. There is evidence that in some cases the sensitivities obtained by this technique are much greater than for other flame methods. For example, the limits of detection for analysis by this fluorescence technique as compared to AAS were for cadmium 0.2 versus 10 µg/l; mercury, 100 versus 500; thallium, 40 versus 200; and zinc, 0.1 versus 5.[7] Nevertheless, atomic fluorescence flame photometry is a relatively new technique and has not been widely used in water analysis.

Emission Spectroscopy

Arc emission spectroscopy has been used for the analysis of many metals in a variety of natural and treated waters, particularly the trace elements. This technique is not sufficiently sensitive for most trace metals and, therefore, before analysis is performed there generally must be preconcentration, such as by evaporation, precipitation, or ion exchange.[13,14] The method can tolerate large quantities of dissolved solids, although macro elements such as sodium, potassium, calcium and magnesium do interfere with the trace elements. However, this can be compensated for by adding these elements to the standards.

In one routine use of this spectroscopic technique in analyzing a large number of samples from surface waters in the United States, all samples were reduced so as to provide the same total quantity of solid material in the dried residue arced in the spectrograph, giving a constant matrix.[14] Thus, since the untreated samples contained variable amounts of dissolved solids, the concentration factor and, hence, the sensitivity of the method varied accordingly. However, in using a chelation-precipitation technique to separate the trace metals from the macro elements, the sensitivity was said to be adjustable at will, since a wide choice of preconcentration factors was available.[23] However, in this case, the precipitation-separation operation was time-consuming.

In many cases, the sensitivity of arc emission spectroscopy without preconcentration is not as good as with other techniques. However, the method is suitable for

240

analyzing a large number of samples. For a variety of surface water systems in the United States, the method of preconcentration by evaporation resulted in the following detection limits in µg/l: Ag and Be, 0.03 to 0.2; Ba, Cr, Cu, Fe, Mn, Mo, Ni, Pb, V, Cd, Co, Sn, 1 to 20; Bi and Sb, 5 to 60; and Zn, 300 to 3500.[14]

Neutron Activation Analysis

Neutron activation analysis is a very sensitive technique for metal analysis, but requires the use of elaborate and expensive equipment not generally available to most laboratories.[14] Examples of sensitivities are gold, 0.0004 µg/l, platinum, 0.1, zinc, .006, cobalt, 0.0006, and copper, 0.003. In addition to the disadvantage of requiring specialized equipment, neutron activation analysis also frequently requires elaborate separation processes for the purpose of eliminating interferences, both prior and subsequent to irradiation, in order to measure separately the radioactivity of each of several elements in a complex mixture.

An example of the application of this technique is the analysis of fifteen lanthanide elements in sea water, the range of concentrations being 0.00013 to 0.0133 µg/l.[13] In this method, iron and uranium were separated from the lanthanides by ion exchange, and precipitation was used for preconcentration. After irradiation with neutrons, gradient elution by ion exchange was used to separate the activated products, and beta counting performed on the separate fractions. It was found that the variation in lanthanide concentration among the oceans was great. For example, the Indian Ocean had total lanthanide concentrations about 100 times that of the deep Central Atlantic Ocean.

Ion Selective Electrodes

Ion selective electrodes are basically potentiometric membrane electrodes systems which operate on a principle similar to that of the glass electrode for the measurement of pH. Their theory and applications for water analysis have been recently reviewed.[1] They are available commercially and measure a wide variety of cations and anions, including the alkali metals, calcium, hardness, nickel, copper, fluoride and other halides, sulfate, nitrate, and sulfide. A list of many of these electrodes is presented in Chapter VI. Their major advantages for use in water analysis are their speed and ability to be used for *in situ* monitoring. However, as with glass pH electrodes,

there are potential interferences which must be recognized
and compensated for when necessary. Similarly, because
they are electrochemical sensors they measure activity
rather than concentration.

The experimental procedure in their use is also quite
similar to that for the glass pH electrode. They are used
with a reference electrode, such as saturated calomel or
silver-silver chloride, and a high impedance potential measuring device, such as a pH meter with a millivolt scale.
A calibration plot of voltage versus the logarithm of concentration of the test ion is made, generally for at least
a 10-fold range of concentration. The voltage for the
test solution is measured, and the test ion concentration
is then calculated from the calibration plot.

Using the calcium electrode, for example, the calibration curve in the absence of interfering ions should
follow the equation

$$E = \text{const.} + RT/2F \ln (A_{Ca^{2+}}) \tag{53}$$

with E as the measured cell voltage, R the gas constant,
T the absolute temperature, F the Faraday, and $A_{Ca^{2+}}$ the
activity of Ca^{2+} species. For dilute solutions, or those
in which the ionic strength is constant, concentration may
be substituted for activity and Equation 53 at 25°C becomes

$$E_{CELL} \text{ (mv)} = \text{const.} + 30 \log [Ca^{++}] \tag{54}$$

However, it should be noted that the theoretical value of
30 mv for the slope of $dE/d \log [Ca^{++}]$ is usually not obtained, and it is necessary to calibrate to be certain of
the value. Also the activity and concentration may not be
equal because the solution is not sufficiently dilute, or
they may not be related in the same way in the test and
calibrating solution. In this case, it is frequently convenient to add electrolyte to all solutions to fix them at
the same ionic strength before calibration and measurement.

Although ideally Equation 53 describes the behavior of
the calcium electrode, other cations may interfere with
its use. For one commercially available calcium elec-
trode,[22] interferences from Mg^{++}, Ba^{++}, H^+, and Na^+ affect
the cell potential according to the equation

$$E_{CELL} = \text{const.} + 30 \log \{(A_{Ca^{2+}}) + 0.014(A_{Ba^{2+}}) +$$

$$0.010(A_{Mg^{2+}}) + 10^5(A_{H^+}) + 10^{-4}(A_{Na^+})\} \tag{55}$$

When a coefficient preceding a particular ion is larger
than unity, the electrode is more sensitive to that ion
than it is to calcium; the reverse also holds. Thus, the

calcium electrode is very sensitive to H^+ and cannot be used at low values of pH. Similarly, Equation 55 indicates that, for example a 0.2M solution of Mg^{++} will give the same reading as a 0.002M solution of Ca^{++}. Thus, low levels of magnesium are readily tolerated and do not interfere with the measurement of intermediate levels of calcium. No electrode should be used without considering such possible interferences.

In using these electrodes to determine the concentration of a test ion, the possibility exists that not all of the soluble species being measured are available as free ions, unassociated with other solutes. Thus, in the use of the fluoride electrode at low pH, it should be noted that because HF is a weak acid with a pH of 3.14, some of the fluoride will be bound to hydrogen ions and not sensed by the electrode; raising the pH will obviate this problem. Complexation of fluoride by aluminum poses a similar difficulty. Calcium and magnesium ions may also be complexed or otherwise bound to ions in natural waters.

All ion selective electrodes have limited sensitivity, usually dependent upon the finite solubility of its sensitive membrane. Thus, this material dissolves at the electrode surface and is sensed by the electrode itself. Although they may require somewhat more care and sophistication in use and interpretation than do pH electrodes, ion selective electrodes are, nevertheless, a powerful tool in water analysis.

Polarography

Polarography is a useful tool in the analysis of metal ions in water because of its high sensitivity and its ability to analyze mixtures, thus reducing the need for prior separations, and to tolerate large quantities of dissolved solids, such as in brines and sea water. A variety of polarographic techniques has been utilized in water analysis and their ranges and advantages reviewed.[18] One study of conventional polarography for the analysis of copper, cadmium, zinc, chromium, and lead in river water, sewage and sewage effluents concluded that, in general, polarography was as sensitive as many good colorimetric analyses, yet was advantageous because it was simpler and separation techniques were not required.[3]

A comparison of the various polarographic techniques indicates that conventional polarography with the dropping mercury electrode is sensitive in the range of 10^{-5}M, pulse polarography around 10^{-7} to 10^{-8}M, cathode-ray polarography better than 10^{-7}M, and anodic stripping polarography sometimes better than 10^{-9}M because of its preconcentration feature.[18] This study also indicated that a combination of cathode-ray and anodic stripping voltammetry appeared to offer excellent sensitivity, one such reported application for zinc analysis in sea water having a sensitivity of 1 µg/l.

An example of the analysis of copper and zinc in river water was given using cathode-ray polarography.[18] Samples of 100 to 500 ml were filtered and evaporated with perchloric acid to destroy organic matter, and diluted with buffer and supporting electrolyte to 5 or 10 ml. The analysis was performed without further separation, with copper at -0.4 volts and zinc at -1.2 volts vs. a mercury pool anode, respectively. In another example, cathode-ray polarography was used to analyze the quality of a central supply of distilled water for a laboratory. After concentrating by factors of 10 to 100, copper in the range of 0.5 to 4.0 µg/l, lead from 0.6 to 8.0 µg/l, and zinc at 5.0 µg/l were determined simultaneously.

In the analysis of river water and brine, samples were evaporated to dryness, while sewage, in addition, was ashed to destroy organics.[3] The residues were then redissolved and additions made of the appropriate electrolyte, buffer, and, where necessary, maxima suppressor, such as polyelectrolyte, and sulfite to reduce oxygen. Using this procedure and conventional polarography for river water samples, to which was added 0.10 mg/l each of copper, cadmium, zinc, chromium, and lead, the recovery was generally complete within 10% and the standard deviation was 4-5%, except for cadmium which was 9%. Similar results were obtained for brine. The authors noted that the polarograph wave height for a well formed wave could generally be measured with a precision of 0.005 mg/l, except for small metal concentrations on the order of 0.01 mg/l. They suggested improvement of sensitivity by extraction with dithizone, followed by its destruction before polarographic analysis.

Because of their applicability to a large number of metals and their advantages cited above, the various polarographic methods present another series of important analytical tools for determining both trace and macro quantities of metals in a wide variety of waters.

Anodic Stripping Voltammetry

Anodic stripping voltammetry (ASV) is an extremely sensitive electroanalytical technique applicable for characterizations of trace metals in aquatic environments. The technique is a combination of a concentration step and an anodic dissolution process (reverse electrolysis or stripping) in which the metal is oxidized and returns to the solution. The former step, either pre-electrolysis, cathodic deposition, or plating is one in which the metal ion (or ions) of interest is reduced by controlled potential electrolysis, either by deposition as a metal onto a solid microelectrode, or by formation of an amalgam with the mercury of a mercury drop or mercury film electrode. ASV, similar to most electrochemical techniques, is therefore a nondestructive method of analysis.

The concentration step is carried out for a definite time under reproducible conditions so that the concentration onto the electrode is quantitative (total plating) or so that a reproducible fraction of the desired component is deposited from solution (partial plating). By controlling the potential during this process, more easily electrolyzed constituents may be separated out. The concentration process is thus performed by applying to the electrodes a potential which is cathodic of the polarographic half-wave potential by some 0.3 to 0.4 volts (where the current has its limiting value) and maintaining this potential for a given period of time under the reproducible conditions of stirring, type and area of microelectrode, and composition of medium. Then, after a short rest period which allows the solution to become quiescent, the stripping process is initiated.

The stripping process is generally performed by some voltammetric scanning procedure which produces a response proportional to the amount of material deposited. The resulting "stripping voltammogram" produces peaks, the heights of which are generally proportional to the concentrations of corresponding electroactive metal ions and the potentials of which are a qualitative indication of the nature of the species present in the solution. On this basis, then, the important characteristics of the peak are its height (peak current, i_p, in microamperes) and the peak potential (E_p, in volts). These characteristics are affected by the type of microelectrode used and by the rate of voltage change (sweep rate, v) employed in the stripping process.

ASV, which is also known as linear potential sweep stripping chronoamperometry, stripping analysis, anodic amalgam voltammetry, and inverse voltammetry, has been successfully applied to the analysis of Ag, Au, Ba, Bi, Cd, Cu, Ga, Ge, Hg, In, K, Mn, Ni, Pb, Pt, Sb, Sn, Tl, and Zn ions.[22a] Its sensitivity (10^{-7} - 10^{-10}M) plus its potential capability of performing *in situ* analysis of free and complexed metal ions make it an ideally suited technique for the analysis of trace metals in natural waters and waste waters.[1a]

Wax impregnated graphite electrodes[19a] and mercury film on graphite electrodes[18a] seem to be most suitable for ASV applications for water pollution measurement.[1a] The mercury film-graphite electrode is composed of an electroplated mercury layer supported on the surface of a paraffin-impregnated graphite rod. This electrode system offers high sensitivity and reproducibility and a ruggedness that makes it very suitable for field work.

The anodic stripping equations for mercury-film electrode have been derived.[21a] The peak current (i_p) is directly proportional to the metal ion concentration in solution

$$i_p = \left[nFAl \; \frac{\phi}{e} v \right] c_R \tag{56}$$

where l is the thickness of the mercury film, ϕ is equal to nF/RT, and e is the base of Napierian logarithms. The peak current is directly proportional to the potential scan rate; in this characteristic the electrode differs from the HMDE, in which the peak current depends on $v^{1/2}$. The peak potential equation for the mercury-film electrode was given as follows

$$E_p = E^O + \frac{e}{\phi} \log \frac{\delta l v \phi}{D} \tag{57}$$

where E^O is the standard electrode potential (which is equal to $E_{1/2}$ in many systems) and δ is the thickness of a boundary diffusion layer at the solution-electrode interface. Experimental variation in δ and l will have little effect on E_p, which generally is cathodic to E^O.

Qualitative and quantitative information on free and combined metal ions and metal-binding ligands and their distribution can be gained by varying the sample pretreatment and the anodic stripping procedures. A flow chart showing various analytical procedures is given in Figure 56. Total metal content can be determined by digestion of the sample before anodic stripping analysis, and filtration separates dissolved from particulate fractions on the basis of variations in particle size.

Metal ions in natural waters may be complexed with simple inorganic ligands such as water, halides, carbonate, and sulfate, or they may be tied up with complex organic ligands such as amino acids, organic acids, vitamins, porphyrins, humic acids, and tannins. It is possible to

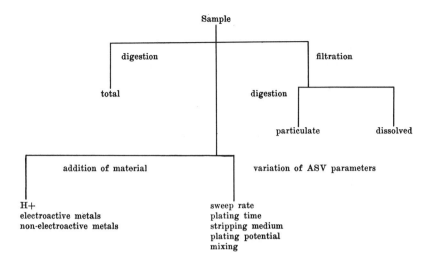

Figure 56. Possible variations in procedures for anodic
          stripping voltammetric determination of metal
          ions in natural waters.

exchange the metal in many complexes by additions of acid
(H^+) or metal cations; the freed metal ions may be determined subsequently by anodic stripping voltammetry
(Figure 56).

The metal-complexing ability of natural waters, which
is a measurable characteristic, depends on the type and
concentration of simple or complex ligands. ASV can be
used advantageously to provide qualitative as well as quantitative information on metal-binding ligands by means of
complexometric titration procedures. In this technique
the stepwise metal-ligand complex formation as well as the
rates of complex formation can be determined.

An example of ASV application for the analysis of Cd,
Pb and Cu in lakes, rivers and waste effluents is shown
in Table 34.

Table 34

Trace Metal Concentrations in Various Natural and Waste Waters Determined by Anodic Stripping Voltammetry

Sample	Location	Concentration (µg/l)				
		Free			Acid Exchangeable	
		Cd	Pb	Cu	Pb	Cu
1	Rouge River 0.5 mile (0.8 km) above Ford Motor Co. turning basin	<0.1	1.1	2	15	27
2	Rouge River turning basin	7.4	2.2	14	37	108
3	Rouge River mouth	1.5	0.4	8	11	19
4	Detroit River, Trenton Channel	0.8	0.4	18	5	28
5	Detroit River, Livingston Channel	1.9	0.4	11	6	29
6	Lake Erie, central basin, surface water	--	--	--	1.8	6.8
7	Lake Erie, central basin, bottom water	0.3	0.07	0.5	1.6	1.8
8	Lake Michigan, Waukegan, Ill.	1.0	0.2	0.8	3.3	19
9	Lake Michigan, Ludington, Mich.	--	--	--	0.6	0.4

REFERENCES

1 . Andelman, J.B. J. Water Pollution Control Federation 40, 1844 (1968).
1a. Allen, H.E., W.R. Matson, and K.H. Mancy. J. Water Pollution Control Federation 42, 573 (1970).

2 . Anderson, N.R., and D.N. Hume. in *Trace Inorganics in Water*, R.F. Gould, Ed., Advances in Chem. Series No. 73, (Washington, D.C.: American Chemical Society, 1968) Chapter 18.

3 . Ballinger, D.G., and T.A. Hartlage. Water Sewage Works 110, 31 (1962).

4 . Biechler, D.G. Anal. Chem. 37, 1054 (1965).

5 . Chau, Y., S. Sim, and Y. Wong. Anal. Chim. Acta 43, 13 (1968).

6 . Chilton, J.M. Anal. Chem. 25, 1274 (1953).

7 . Ellis, D.W., and D.R. Demers. in *Trace Inorganics in Water*, R.F. Gould, Ed., Advances in Chem. Series No. 73, (Washington D.C.: American Chemical Society, 1968) Chapter 20.

8 . Fair, G.M., J.C. Geyer, and D.A. Okun. *Water and Waste Engineering*, Vol. 2 (New York: John Wiley and Sons, Inc., 1968).

9 . Fishman, M.J., and M.R. Midgett. in *Trace Inorganics in Water*, R.F. Gould, Ed., Advances in Chem. Series No. 73, (Washington, D.C.: American Chemical Society, 1968) Chapter 12.

10 . Garrels, R.M., and M.E. Thompson. Am. J. Sci. 260, 57 (1962).

11 . Greenhalgh, R., J.P. Riley, and M. Tongudai. Anal. Chim. Acta, 36, 439 (1966).

12 . Hem, J.D. U.S. Geological Survey Water-Supply Paper 1459-B, Washington, D.C., 1960.

13 . Hogdahl, O.T., and S. Melsom. in *Trace Inorganics in Water*, R.F. Gould, Ed., Advances in Chem. Series No. 73, (Washington, D.C.: American Chemical Society, 1968) Chapter 19.

14 . Hume, D.N. in *Equilibrium Concepts in Natural Water Systems*, R.F. Gould, Ed., Advances in Chem. Series No. 67, (Washington, D.C.: American Chemical Society, 1967) Chapter 2.

15 . Jones, J.L., and R.D. Eddy. Anal. Chim. Acta 43, 165 (1968).

16 . Kahn, H.L. in *Trace Inorganics in Water*, R.F. Gould, Ed., Advances in Chem. Series No. 73, (Washington, D.C.: American Chemical Society, 1968) Chapter 11.

17 . Kroner, R.C., and J.F. Kopp. J. Am. Water Works Assoc., 57, 150 (1965).

18 . Maienthal, E.J., and J.K. Taylor. in *Trace Inorganics in Water*, R.F. Gould, Ed., Advances in Chem. Series No. 73, (Washington, D.C.: American Chemical Society, 1968) Chapter 10.

18a. Matson, W.R., D.K. Roe, and D.E. Caritt. Anal. Chem. 37, 1594 (1965).

19 . Morris, J.C., and W. Stumm. in *Equilibrium Concepts in Natural Water Systems*, R.F. Gould, Ed., Advances in Chem. Series No. 67, (Washington, D.C.: American Chemical Society, 1967) Chapter 13.

19a. Perone, S.P., and W.J. Kretlow. Anal. Chem. **37**, 968 (1965).

20 . Platte, J.A. in *Trace Inorganics in Water*, R.F. Gould, Ed., Advances in Chem. Series No. 73, (Washington, D.C.: American Chemical Society, 1968) Chapter 14.

21 . Riley, J.P., and M. Tongudai. Chem. Geol. **1**, 291 (1966).

21a. Roe, D.K., and J.E.A. Toni. Anal. Chem. **37**, 1503 (1965).

22 . Ross, J.W. Science **156**, 1378 (1967).

22a. Shain, I. "Stripping Voltammetry," in *Treatise in Analytical Chemistry* Part I, Vol. 4 (New York: John Wiley and Sons, Inc., 1964).

23 . Silvey, W.D., and R. Brennan. Anal. Chem. **34**, 784 (1962).

24 . Sulcek, Z., P. Povondra, and R. Stangl. Chemist-Analyst **55**, 36 (1966).

25 . Tenny, A.M. Perkin-Elmer Instr. News **18**, 1 (1967).

26 . Thompson, M.E. Science **153**, 866 (1966).

27 . Thompson, M.E., and J.W. Ross. Science **154**, 1643 (1966).

Chapter XII

Analysis for Inorganic Anions

This chapter is concerned with instrumental analysis of non-metal inorganic species. Nitrogen and phosphorus compounds, commonly classified as "nutrients," will be discussed separately.

Separation Techniques

Separation and concentration of inorganic anions in natural and waste water samples are carried out by a variety of techniques. Evaporation, precipitation, ion exchange and partial freezing are particularly applicable. Chromatographic techniques have also been used, e.g., ortho-, pyro-, tri-, trimeta-, tetrameta-, and polyphosphates have been developed[16,36] with strongly basic anion exchange resins. Sulfate, sulfite, thiosulfate and sulfide ions can be separated by anion exchange chromatography using the gradient elution technique with an elution solution of nitrates.[17] A more detailed discussion of the possible applications of anion exchange for such analytical separations has been given by Inczedy.[18]

Spectrophotometric Techniques

Many highly sensitive absorption spectrophotometric techniques are used for the analysis of inorganic anions. Some of these techniques are based on the displacement of a ligand (usually colored) in a metal complex or chelate. The analysis for fluoride ions by displacement of a chelating dye anion from a zirconium complex is a typical example of this technique. The displaced dye differs sufficiently in color from its zirconium chelate to permit determination of the fluoride ion concentration as a function of the change in color of the solution.[5]

Numerous indirect spectrophotometric methods have been developed.[6] Heteropoly chemistry appears to offer important advances in analyses for phosphates, silicates and arsenates.[8] Reviews in *Anal. Chem.* give an exhaustive survey on the subject matter.[6]

Indirect UV spectrophotometry and atomic absorption methods have been developed for phosphates and silicates.[16] These techniques are based on the selective extraction of molybdophosphoric acid and molybdosilicic acid, followed by ultraviolet molecular absorption spectrophotometry and/or atomic absorption spectrophotometry. The molybdophosphoric and molybdosilicic acids are formed in acidic solution by addition of excess molybdate reagent. Molybdophosphoric acid is extracted with diethyl ether from an aqueous solution which is approximately 1M in hydrochloric acid. After adjusting the hydrochloric acid concentration of the aqueous phase to approximately 2M, the molybdosilicic acid is extracted with 5:1 diethyl ether-pentanol solution. The extracts of molybdophosphoric and molybdosilicic acids are subjected to acidic washings to remove excess molybdate. Each extract is then contacted with a basic buffer solution to strip the heteropoly acid from the organic phase. The molybdate resulting from the decomposition of the heteropoly acid in the basic solution is then determined either by measurement of the absorbance at 230 mμ using ultraviolet spectrophotometry or by measurement of absorbance at the 313.3 mμ resonance line of molybdenum by atomic absorption spectrophotometry. Optimum concentration ranges are approximately 0.1-0.4 mg/l of phosphorus or silicon for indirect ultraviolet spectrophotometry and 0.4-1.2 mg/l for indirect atomic absorption spectrophotometry.

Electrochemical Techniques

Electrochemical methods of analysis offer considerable promise for selective and specific measurements in waste waters. Direct potentiometric techniques may be used for the analysis of chlorides or sulfides. Silver electrodes, coated with a layer of halides or sulfides, are available commercially for determination of chlorides, bromides, iodides, and sulfides at concentrations corresponding to the solubilities of the respective silver salts, as given by the following expressions:

$$E_{cell} = (E^0_{Ag^+,Ag} - E_{ref}) + \frac{RT}{nF} \ln K_{sp} - \frac{RT}{nF} \ln a_{Cl^-} \quad (58)$$

and

$$E_{cell} = (E^0_{Ag^+,Ag} - E_{ref}) + \frac{RT}{nF} \ln K_{sp} - \frac{RT}{nF} \ln a_{S^{2-}} \quad (59)$$

where K_{sp} refers to the solubility products of AgCl and Ag_2S, respectively, in Equations 58 and 59. Limitations on the use of such electrode systems are imposed by interferences from other potential-determining ions, by the problem of elimination of liquid junction potentials, and by the difficulty of satisfactorily resolving single ion activity coefficients.

Major developments in electrochemical analysis for anions have occurred in the area of ion-selective electrodes. Such electrode systems are primarily solid state or precipitate, ion exchange membrane electrodes. Pungor and co-workers[27] have developed anion selective membrane electrodes by impregnating silicone gum rubber membranes with specific insoluble salts. Electrodes which are selective for Cl^-, Br^-, I^-, S^{2-}, SO_4^{-2} and F^- have been reported.[26,28] Pungor[26] has reviewed the literature of ion selective membrane electrodes and discussed the preparation and characterization of the membrane. Response time, memory effects, and detection limits up to $10^{-4} M$ have been discussed. Detailed studies on the sensitivity and selectivity of iodide, bromide and chloride electrodes have been reported by Rechnitz, et $al.$[30] The electrodes were prepared by incorporating AgI, AgBr, and AgCl into silicone rubber matrices. Selectivity ratios for the various electrodes were reported with response times $(t_{1/2})$ of 8, 14, and 20 seconds, respectively. The lower limits of response (Equation 60) were $10^{-7}M$, I^-, $5 \times 10^{-4}M$ Cl^-, and $7 \times 10^{-5}M$ Br^-.

$$E = \text{const.} + RT/F \ln \{a_{A^-} + K a_{B^-}\} \quad (60)$$

where a_{A^-} and a_{B^-} are the activities of A^+ and B^+ ions, respectively, and K is a selectively constant which expresses the relative sensitivities of the glass electrode for ions B^+ and A^+.

Electrode systems prepared by incorporating metal oxides and hydroxides[15] or ion exchange resins[27] in polymeric membranes also have been reported. Solid state ion exchange membrane electrodes for fluoride ion determinations have been reported.[13] This type of membrane electrode is made of single crystals of LaF_3, NdF_3, and PrF_3 and is reported to be sensitive to fluoride ion concentrations as low as $10^{-5}M$.

Anion selective membrane electrodes can be used for direct potentiometry or for potentiometric titrations. For example, Lingane[23] used a commercially available fluoride electrode for the potentiometric titrations of Th^{4+}, La^{3+} and Ca^{2+} ions. Best results were obtained with $La(NO_3)_3$. The equivalence point potential was determined within ±2mV and an accuracy better than 0.1% was noted in neutral, unbuffered solutions.

The fluoride ion membrane electrode finds its widest application in the analysis of fluoridated water supplies.[2,20] Free fluoride ions are determined in water samples buffered to pH 5.2 by an acetate buffer. Total fluorides are determined after the addition of a masking agent, *e.g.*, citrate, which forms stable, nonmobile complexes with aluminum and measuring fluorides of pH 5.2. The addition of a masking agent is necessary to release the bound fluorides, *e.g.*, $(AlF_6)^{3-}$. In this procedure, the fluoride ion potentiometric membrane electrode is used to provide quantitative information on free and complexed fluorides. This should be highly significant in assessing the effect of free vis-a-vis complex fluorides on the prevention and control of dental caries.

Similarly, the sulfide ion potentiometric membrane electrode can be used to characterize free sulfides in natural and waste waters.[9] Potentiometric measurements of S^{2-} can be related to the concentrations of H_2S, HS^-, and S_T (analytical concentration of free sulfides) by the following equilibrium relationship

$$\log [S_T] = \log [S^{2-}] + \log \left(\frac{[H^+]^2}{K_1 K_2} + \frac{H^+}{K_1} + 1\right) \quad (61)$$

$$\log [H_2S] = \log [S^{2-}] + \log \frac{[H^+]}{K_1 K_2} \quad (62)$$

$$\log [HS^-] = \log [S^{2-}] + \log \frac{[H^+]}{K_1} \quad (63)$$

where K_1 and K_2 are the acidity constants for the weak acid H_2S and activity coefficients are assumed to be unity.

Cyanide nitrates and chloride ions, in natural and waste waters, may be determined also by potentiometric membrane electrodes.[22,24] These applications are subject to the same general limitations on sensitivity and selectivity as those for previously mentioned membrane electrodes. It is important, however, to establish the electrode performance characteristics in any given water prior to its application. For example, the nitrate electrode is quite accurate and precise (1-5% error) in pure nitrate solutions, and the lower limit of sensitivity

is about 0.66 ppm NO_3^- which is sufficient to give meaningful electrode response at any nitrate level in surface and
underground waters in the United States. Nevertheless,
the electrode's liquid ion exchanger deteriorates sufficiently to cause significant errors after three to four
weeks of operation, after which the electrode must be re-
juvenated and recalibrated.

One of the main problems with the nitrate electrode is
the effect of interferences by Cl^- and HCO_3^-, commonly
found in all types of waters. The selectivity coefficients
for Cl^- and HCO_3^- ions have been determined[22,33] and were
found to vary with nitrate ion activity. For U.S. tap
water, as reported by USGS[10] the median chloride ion activity is $1X10^{-4}M$ and the median nitrate ion activity is
$1X10^{-5}M$. Under these conditions there exists about 75%
error in nitrate measurement by the membrane electrode.
Errors of similar magnitude exist for HCO_3^- interference.

It is evident that errors due to Cl^- and HCO_3^- interferences are far from acceptable, and methods have been investigated for removing these ions from the test solution
or to correct for their presence. This can be done by
manual or automatic compensation of the nitrate electrode
response based on (a) the selectivity coefficient of the
interfering ion, (b) the activity of the interfering ion,
and (c) the nitrate ion activity. Removal of Cl^- and HCO_3^-
from the test solution may be achieved by precipitation as
silver and aluminum salts.[25] This is considered, however,
to be a difficult procedure to apply to samples of unknown
composition. A more adequate procedure is that of calibrating the nitrate·electrode by standard additions to the
same test solution provided that changes in selectivity
coefficient with changes in nitrate ion activity are taken
into consideration.

Nitrogen Compounds

The presence of combined nitrogen (ammonia, nitrites
and nitrates) NH_3, NO_2^- and NO_3^- in natural waters may
result from the degradation of organic nitrogenous compounds or may be entirely of inorganic origin.

The most widely used method for analysis for ammonia is
the Nesslerization reaction. The test is based on the development of a yellow-brown (colloidal) color on addition
of Nessler's reagent to an ammonia solution. The standard method[3] and the ASTM reference test[1] recommend the
separation of the ammonia from the sample by distillation
prior to the Nesslerization reaction. Direct Nesslerization is most often preferred, however, for rapid routine
determinations.

For certain industrial waste waters, it is often desirable to distinguish between "free" ammonia and "fixed"
ammonia. The former is estimated by a straightforward
distillation; the residual liquor is then treated with an
alkali (e.g., sodium carbonate, magnesium oxide or caustic
soda) and distilled to determine fixed ammonia. Certain
substances interfere with both the direct Nesslerization
and distillation methods, e.g., glycine, urea, glutamic
acid, acetamides and hydrazines.

The standard method[1],[3] for nitrites in water is based
on forming a diazonium compound by the diazotization or
sulfonilic acid by nitrite under strongly acidic conditions and coupling with α-naphthylamine hydrochloride to
produce a reddish-purple color. Spectrometric measurement of the color of the azo dye is performed at 520 mµ,
or visual comparison with standards may be used. This
method is sometimes known as the Griess-Ilosvay method.[31]
A frequently used alternative procedure for nitrite is
based on formation of a yellowish-brown dye by the reaction
of nitrite in acid solution with p-phenylene diamine.[21]

The Griess-Ilosvay method is most suitable for low nitrite concentrations, e.g., below 2.0 mg/l. Another colorimetric procedure which is more suitable for high concentrations involves reaction of the nitrites with p-phenylene
diamine in an acid solution to form a yellowish-brown dye.
Chlorides in concentration below 500 mg/l have no effect
on either of the two procedures. High chloride concentrations (10,000 mg/l) interfere more with the Griess-Ilosvay
test. More distinct, pH-independent, color development
is achieved[34] by replacing the sulfanilic acid with
sulfanilamide and α-naphthylamine with 1-naphthyl-ethylene
diamine dihydrochloride within the Griess-Ilosvay chloric
acid and the diazonium salt is then coupled with the
diamine to give a stable red azo-dye.
the diamine to give a stable red azo-dye.

An excellent review of approximately 52 spectrophotometric methods for nitrite has been reported by Sawicki,
et al.[32] The authors critically evaluate the sensitivity,
color stability, conformity to Beer's law, simplicity and
precision of a variety of methods.

Colorimetry, UV spectrometry and polarography have frequently been used for nitrate determinations in natural
waters and waste effluents. The phenoldisulfonic acid
method and the brucine method are two colorimetric procedures more frequently used. In the former test, color
development is based on the reaction between phenol disul-
fonic acid and nitrates in sulfuric acid solution to give
a nitro-derivative which causes a yellow coloration when

256

the solution is made alkaline; the intensity of the color is measured at 470 mµ. Nitrite ion interferes with the test in proportion to its concentration in the sample. Various inorganic ions have certain concentrations which cause interference.[1] Small amounts of chlorides do not interfere but nitrites should be removed with sodium azide.[1]

An alternate test for nitrates involves reaction of a brucine solution in glacial acetic acid with nitrates and acidification with dilute H_2SO_4. As the color intensity changes with time, it is necessary to develop the color of standards and samples simultaneously and compare maximum color intensity. Chlorides above 1,000 mg/l interfere with color development. Nitrites, if present, should be separately estimated and an appropriate correction applied. A salt-masking technique which renders the test applicable to sea water and brackish water has been proposed by Jenkins.[19]

Nitrate analysis by reduction to ammonia, which is then detected by Nesslerization, has been reported by several authors. The procedure is based on expelling all ammonia from the water sample, followed by reduction of nitrogen (NO_2^- and BI_3^-) by means of (a) aluminum foil in alkaline NaOH solution, (b) zinc-copper couple in acetic acid solution, (c) Dervado's alloy hydrazine,[7] and (d) alkaline ferrous sulfates. The ammonia produced may be separated by steam distillation and estimated in the distillate by Nesslerization. Various procedures have been proposed to minimize interferences due to nitrites and chlorides.

Nitrate analysis by reduction to nitrites which are then detected by the Griess-Ilsovay method has been applied to both natural waters and waste waters.[11] Controlled reduction of nitrates to nitrites is accomplished with zinc powder in acid solution.

Ultraviolet analysis for nitrates offers the advantage of freedom from chloride interferences and a variety of other inorganic ions. However, dissolved organic compounds, nitrites, hexavalent chromium, and surfactants interfere with this procedure. The test is based on measuring UV adsorption spectra of the filtered, acidified sample at 220 mµ. Measurements follow Beer's Law up to 11 Mg N/l. Interference of dissolved organics is estimated by doing a second measurement at 275 mµ, a wave length at which nitrates do not absorb.

Simultaneous determination of nitrates, nitrites and sulfates in water samples by infrared techniques has been reported.[19] The test is based on concentrating the sample by ion exchange and removing of phosphates, carbonates and organic matter. This is followed by separation by freeze drying the aqueous solution in the presence of KBr; the infrared spectrum is determined in the resulting KBr disk.

The polarographic analysis of nitrate is based on diffusion current measurements in an acid solution at -1.2 volts vs. the S.C.E. Nitrites, phosphates, ferric iron and fluorides interfere with the test. Procedures to minimize interferences have been prescribed. The polarographic test offers the advantage of being adaptable to continuous monitoring.[1,3,14]

Phosphorus Compounds

Phosphorus may be present in industrial waste effluents either as inorganic phosphates (ortho-, meta-, or polyphosphates) or in organic combination. The most common analytical method for inorganic phosphorus is based on the colorimetric determination of the phosphomolybdenum blue complex.[1,3] The test is specific for orthophosphates. Polyphosphates and metaphosphates are then estimated as the difference between total phosphates (hydrolyzed samples) and orthophosphates (non-hydrolyzed samples).

Orthophosphates react with ammonia molybdate in acid medium to form the phosphomolybdic acid complex, which when reduced yields the molybdenum blue color which may be determined colorimetrically. The sensitivity of the test is largely dependent on the method of extractions and reduction of the phosphomolybdic acid, aminonaphtholsulfonic acid,[3] stannous chloride (Deniges Method),[4] metal sulfites (Tschopp reagent),[35] and ascorbic acid[12] have been used in the reduction step. The stannous chloride method is considered most sensitive and best suited for lower ranges of phosphate concentration.

A number of substances have been reported to interfere with the phosphate determination.[3] Arsenic, germanium, sulfides, and soluble iron above 0.1 mg cause direct interferences. Tannins, lignins and hexavalent chromium will cause errors only at low phosphate concentrations.

REFERENCES

1. American Society of Testing Material, "Manual on Industrial Water and Industrial Waste Water," in *American Society for Testing Materials*, 2nd ed. (1964).

2. Analytical Reference Service, Environmental Control Administration, U.S.P.H.S., "Water Fluoride No. 3," Public Health Service Publication No. 1895 (1969).

3. APHA, AWWA, and WPCF. "Standard Methods - Water and Waste-Water," in *American Public Health Assoc.*, 12th ed. (1965).

4. Association of British Chemical Manufacturers and Society of Analytical Chemistry Joint Committee. Anal. Chem. 83, 50 (1958).

5. Belcher, R., M.A. Leonard, and T.S. West. Talanta 2, 92 (1954).

6. Boltz, D.F., and M.G. Mellon. Anal. Chem. 40, 255R (1968).

7. Citron, I., H. Tsi, R.A. Day, and A.L. Underwood. Talanta 8, 798 (1961).

8. Crouch, S.R., and H.V. Malmstadt. Anal. Chem. 39, 1084 (1967).

9. Demirjian, Y.A., and K.H. Mancy. Research report, The University of Michigan, School of Public Health, Ann Arbor, Michigan (1969).

10. Durfor, C.N., and E. Becker. "Public Water Supplies of the 100 Largest Cities in the United States, 1962," in *U.S. Geological Survey Water Supply Paper 1812* (1964).

11. Edwards, G.P., J.P. Pfafflin, L.H. Schwartz, and P.M. Lauren. J. Water Pollution Control Federation 34, 1112 (1962).

12. Fogg, D.N., and N.T. Wilkinson. Anal. Chem. 83, 406 (1958).

13. Frant, M.S., and J.W. Ross, Science 154, 1553 (1966).

14. Frasier, R.E. J. Am. Water Works Assoc. 55, 624 (1963).

15. Geyer, R., and W. Syring. Z. Chem. 6, 92 (1966).

16. Grande, J.A., and J. Beukenkamp. Anal. Chem. 28, 1497 (1951).

17. Hurford, T.R., and D.F. Boltz. "Abstracts of Papers," 154th American Chemical Society Meeting, Chicago (1967).

18. Inczedy, J. *Analytical Applications of Ion Exchangers,* Pergamon Press (1966) p 151.

19. Jenkins, D., and L.L. Medsker. Anal. Chem. 36, 610 (1964).

20. Kelada, N.F., N.I. McClelland, and K.H. Mancy. "Monitoring of Fluorides in Water Supplies," Progress report, National Sanitation Foundation, Ann Arbor, Michigan (1970).

21. Klein, L.J. "Observations on the Determination of Nitrites in Sewage Effluents," in J. Inst. Sewage Pur. 2, 153 (1950).

22. Langmuir, R., and R.L. Jacobsen. Environ. Science and Tech. 4, 10 (1970).
23. Lingane, J.J. Anal. Chem. 39, 881 (1967).
24. Orion Research Inc. Newsletter, "Research Bibliography on Orion Specific Ion Electrodes" (Cambridge, Mass.: Orion Research Inc., 1970).
25. Paul, J.L., and R.M. Carlson. J. Agriculture and Food Chemistry 15, 5 (1968).
26. Pungor, E., and J. Havas. Acta Chim. Acad. Sci Hung. 50, 77.
27. Pungor, E., K. Toth, and J. Havas. Acta Chim. Acad. Sci. Hung. 41, 239 (1964).
28. Pungor, E., K. Toth, and J. Havas. Acta Chim. Acad. Sci. Hung. 48, 17 (1966).
29. Rechnitz, G.A., and M.R. Kresz. Anal. Chem. 38, 1786 (1966).
30. Rechnitz, G.A., M.R. Kresz, and S.B. Zamochnich. Anal. Chem. 38, 973 (1966).
31. Rider, B.F., and M.F. Mellon. Ind. Eng. Anal. Chem. 18, 96 (1946).
32. Sawicki, E., T.W. Stanley, J. Pfaff, and A.D. Amico. Talanta 10, 641 (1963).
33. Schwenk, T., N.I. McClelland, and K.H. Mancy. "Monitoring of Nitrates in Water Supplies," Progress report, National Sanitation Foundation, Ann Arbor, Michigan (1970).
34. Shinn, M.B. Ind. Eng. Anal. Chem. 13, 33 (1941).
35. Tschopp, Ernst, and Emilia Tschopp. Helv. Chim. Acta. 15, 793 (1932).
36. Weiker, W. Anal. Chem. 185, 457 (1962).

Chapter XIII

This discussion of analyses for dissolved gases in natural and waste waters is concerned only with elemental gases such as molecular nitrogen and oxygen. Analyses for important volatile inorganic weak acids and bases, such as H_2S, CO_2, and NH_3, have been discussed in earlier sections.

Separation Techniques

Dissolved gases in waste effluents usually may be separated rather readily by vacuum degasification or by one or more of various stripping techniques. Stripping is essentially a gas-liquid extraction procedure in which an inert carrier gas is bubbled through a sample to carry off the dissolved gases for further separation, concentration, or detection. Gas transfer efficiency in such systems is dependent on the gas-liquid interfacial area and on the degree of mixing.

Gas-exchange separation can be carried out as either a batch or a continuous flow process. In one design, a continuous mixed stream of sample and carrier gas (nitrogen or hydrogen) is forced through an aspirator nozzle under 50 pounds of pressure.[9] In another design, the dissolved gases are stripped from the test solution by multiple spinning discs rotating at high speed.[36,43] Detection of the stripped gases in the stream of carrier gas may be done by measurement of paramagnetic susceptibility, thermal conductance, etc.[24]

The gas stripped from a waste water sample may be separated into its various components by gas chromatography.[43] Several modifications of this technique have been reported.[33,34,36,37,38] By choosing appropriate detectors, it is usually possible to analyze simultaneously for almost all gases of interest in a water or waste water sample.

Dissolved Oxygen

The following detailed discussion of the instrumental methods of analysis for dissolved oxygen in natural and waste waters will serve to exemplify various techniques, some of which are applicable to the analysis of dissolved gases. Needless to say, dissolved oxygen measurement is a significant indicator of water pollution and a vital parameter in biological waste treatment processes.

Physicochemical properties of dissolved oxygen

Surface waters obtain their oxygen supply mainly from the atmosphere and from the photosynthesis of green aquatic flora. Photosynthetic oxygen production by the aquatic flora may result in an oxygen oversaturation of bodies of water.

The solubility of atmospheric oxygen in water depends primarily on pressure, temperature, and salt content. Atmospheric oxygen solubility in water closely follows the laws of ideal gases, as shown in Figure 57, and can be calculated from Henry's law.

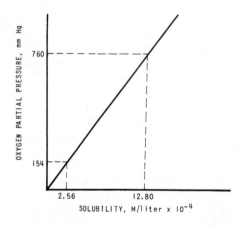

Figure 57. Solubility of oxygen in water at 25°C.

$$P_{O_2} = Ha = H\gamma C \qquad (64)$$

where H is Henry's law constant, γ is the activity coefficient and a and C are the activity and concentration of molecular oxygen in the molar scale, respectively. The activity, defined as the relative fugacity or the ratio of the fugacity in any given state to the fugacity in some standard state, is used in this study as a direct measure of the difference between the partial modal free energies ($\overline{G}$), or the chemical potentials (μ) of dissolved oxygen, or the chosen and standard states

$$\overline{G} - \overline{G}^0 = \mu - \mu^0 = RT \ln a \qquad (65)$$

An aqueous solution of molecular oxygen in equilibrium with one atmosphere of air ($PO_2 = 154$ mm Hg) is sufficiently dilute, and it is assumed that $\gamma = 1.0$, hence $a = C$. If, however, a strong electrolyte is introduced to the aqueous phase, it will exert a "salting out" effect, and as a result $\gamma > 1.0$ and $a > C$. The increase in the activity coefficient is usually expressed by the following equation

$$\ln \gamma = K_s I \qquad (66)$$

where K_s is the salting out coefficient and I is the ionic strength.

Alternately, if an organic solvent, *e.g.* ethanol, is introduced into the aqueous phase, it will exert a "salting in" effect, and as a result $\gamma < 1.0$.

As with most gases, the solubility of oxygen in water decreases with increasing temperature. At 1 atmosphere (atm) of air, oxygen solubility in water at $0^{\circ}C$ is 14.74 mg/liter, whereas at $35^{\circ}C$ it is 7.03 mg/liter. This effect of temperature was found to conform to the following equation.[29,32]

$$C = \alpha + \beta_1 T_c + \beta_2 T_c^2 + \beta_3 T_c^3 \qquad (67)$$

where T_c = temperature, $^{\circ}C$, and α, β_1, β_2, and β_3 = constants.

An easier formula for the interpolation of DO saturation values at various temperatures is:

$$\frac{d (\ln C)}{d (1/T)} = \frac{\Delta H}{R} \qquad (68)$$

where T = absolute temperature, $^{\circ}K$; ΔH = heat of solution, cal/M; and R = gas constant = 1.99 cal/M/degree. Although the heat of solution, ΔH, is not independent of temperature, it normally varies slightly so that use of log C and $1/T$ as variables provides a linear relationship. Integration of Equation 68 gives

$$\log \frac{C_2}{C_1} = \frac{\Delta H}{2.3R} \left(\frac{1}{T_1} - \frac{1}{T_2}\right) \qquad (69)$$

Instrumental modifications of Winkler test

The oldest and one of the most popular methods for the analysis of dissolved oxygen is the Winkler test. Originally reported about 75 years ago, the Winkler procedure possesses most attributes of basic soundness and sensitivity. Improved by variations in equipment and techniques and aided by modern instrumentation, this test is still the basis for the majority of titrimetric procedures for dissolved oxygen. The test is based on the quantitative oxidation of manganese(II) to manganese(IV) under alkaline conditions. This is followed by the oxidation of iodide by the manganese(IV) under acid conditions. The iodine so released is then titrated with thiosulfate in the presence of a starch indicator. The reported precision of the standard Winkler test is ±0.1 mg/l of dissolved oxygen.[1]

Electrochemical detection of end point in the Winkler titration increases the sensitivity, accuracy and precision of the technique. This can be achieved by potentiometric end point detection of the iodometric titration using a platinum wire electrode and a saturated calomel reference electrode. The stoichiometric end point is characterized by the inflection point in the titration curve.

"Dead-Stop" end point methods can also improve significantly the Winkler procedure. This special type of amperometric technique, noted for its simplicity, is probably the most precise and sensitive method for titrating iodine with thiosulfate. In this technique a small potential difference (15 to 400 mv, depending on the sensitivity required) is set up between two smooth platinum electrodes in the solution to be titrated (Figure 58).

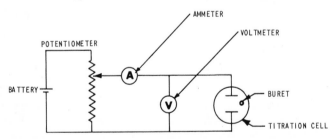

Figure 58. Dead-stop end point circuit.

The diffusion current is measured during the course of the titration. No attempt is made to control the potential of either electrode during titration; only the potential difference is controlled. The end point is indicated by disappearance of the current flowing in the cell as shown in Figure 59.

The course of the titration is as follows: As long as free iodine remains in the solution, the chief electrode reaction, under the influence of the applied voltage, will be the oxidation of iodide to iodine at the anode and the reverse process at the cathode. At the end point and when the free iodine is removed, the iodine-to-iodide reaction can no longer occur and the cell current will come to a "dead stop." Since the thiosulfate/tetrathionate reaction is highly irreversible and proceeds at only a minute rate under the influence of the applied voltage, no detectable

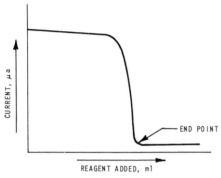

Figure 59. Dead-stop end point titration curve.

current is observed at and beyond the end point. Ordinarily, the end point is so easily detected that there is no need for a graphical estimation of its position. By using sensitive current-measuring devices the end point can be established to an accuracy of ±0.01 microgram iodine in a 100-milliliter sample, or 1 part in 10 billion.

Coulometric titration

Coulometric titration procedures can also be used to improve the Winkler titration. The unique aspect is that in coulometric titrations the titrant (or reagent) is electrically generated at the surface of electrodes immersed in solution. Coulometric titrations follow Faraday's law, expressed mathematically as follows:

$$G = \frac{qM}{nF} \tag{70}$$

where G is weight of substance (iodine), M is molecular weight of substance, q is number of coulombs passed, n is number of equivalents per molecular weight (or the number of electrons in the half-cell reaction), and F is the Faraday.

265

Most coulometric titrations are performed at constant current to simplify the instrumentation and to produce an answer directly proportional to time as shown by the following equation:

$$q = i \int_0^t dt = it \qquad (71)$$

where t is time and i is current.

A typical constant-current coulometer is illustrated in Figure 60. The procedure for coulometric titration of DO consists of the addition of $MnSO_4$, $KOH+KI$, H_2SO_4, and an

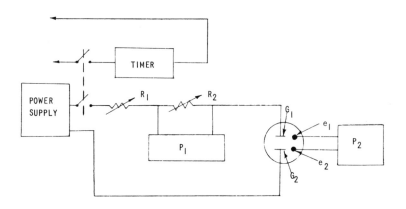

R_1, R_2 = Resistors
P_1 = Potentiometer for current measurement
P_2 = Potentiometer for end point detection
e_1, e_2 = Monitoring electrodes
G_1 = Generating electrode
G_2 = Auxiliary electrode

Figure 60. Constant-current coulometric circuit.

excess of $Na_2S_2O_3$ standard solution successively to the test solution.[16] The iodine formed electrolytically reacts with the residual thiosulfate in solution. The electrolytic current is held constant by varying the potential, and the equivalent point is conveniently detected by the dead-stop end point method. The concentration of DO is calculated from Equations 70 and 71. Since i is constant, by substitution of time of electrolysis, t, in Equation 70, q (the number of coulombs) is determined. By

substitution of q in Equation 70 the weight, g, of DO in
the sample can be calculated.

The coulometric titration method is very accurate, to
within 0.02 µg/liter. It has a distinct advantage in that
because the titrant is generated in solution, errors
caused by contact with air, as in conventional titrations,
are eliminated.

Gas exchange methods

The main difficulty encountered in the determination of
DO in natural and waste waters is the effect of interfering substances present in the sample. One way to avoid
this effect is to "pull out" oxygen molecules from the
test solution and determine their quantity in the gas
phase. This is the basic principle of the gas-exchange
oxygen analyzers. The method used is to exchange oxygen
in the test solution with a carrier gas and then measure
the oxygen in the carrier gas. A typical oxygen analyzer
of this type is composed of two parts, the gas exchange
unit and the oxygen-detecting unit.

(A) Gas exchange unit

A carrier gas must be inert (*e.g.*, nitrogen or hydrogen) so that it does not react with oxygen under test conditions or activate the sensing element in the oxygen-
detection unit. The gas exchange unit is usually composed
of an aspirator followed by a water-gas separator, as
shown in Figure 61. In the aspirator unit, the water
sample with the carrier gas is forced through a nozzle at
about 50 pounds pressure. Intimate mixture of the gas
and water results in the transfer of oxygen from the
aqueous phase to the gas stream (Henry's law). The extent
of the gas transfer in this process depends on the efficiency of the gas exchange unit. It is doubtful, however,
that an instantaneous equilibrium partition of oxygen between the incoming sample and the carrier gas will take
place; therefore, water and gas flow rates must be kept
uniform to maintain a certain degree of gas exchange.[38]

The carrier gas leaving the exchange unit is passed
through the oxygen-detecting unit and then is either
vented to the atmosphere[38] or recirculated through the
exchange unit.[9,20,21,22] In the latter technique, after
the carrier gas has been circulated a few times, equilibrium conditions are attained. For all practical purposes,
at equilibrium conditions, the oxygen in the solution is
assumed to be completely replaced by the carrier gas.

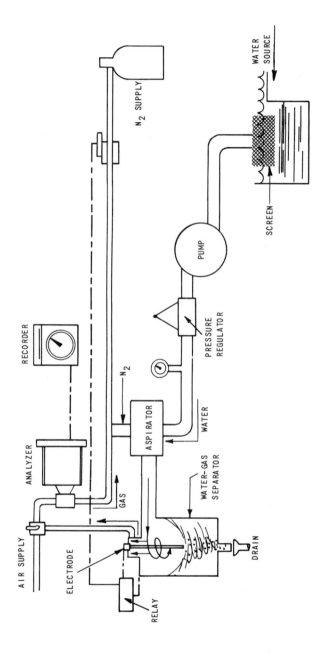

Figure 61. Gas—exchange oxygen analyzer.

(B) Oxygen-detecting units

Three methods are commonly used to determine the oxygen content in the carrier gas. (1) Oxygen-sensitive galvanic cell: Oxygen-sensitive galvanic cells are composed of a relatively basic metal anode, *e.g.*, zinc, cadmium, or lead, a relatively noble cathode, *e.g.*, gold, silver, or nickel (or carbon), and a highly conductive electrolyte solution, *e.g.*, potassium hydroxide. The carrier gas containing oxygen is bubbled through the electrolyte solution by means of diffusers. The oxygen goes into solution and is subsequently depolarized at the cathode; an equivalent amount of current is thus generated.[13,18] The generated current, directly proportional to the oxygen content in the original water or waste solution, is usually measured by a microammeter calibrated in terms of oxygen.

An example of the reactions in an oxygen-sensitive galvanic cell made of a zinc-carbon couple in a $1M$ potassium hydroxide solution is as follows:

Anodic reaction (zinc electrode)

$$Zn \longrightarrow Zn^{2+} + 2e \qquad (72)$$

$$Zn^{2+} + 4\ OH^- \longrightarrow ZnO_2^{2-} + 2\ H_2O \qquad (73)$$

Cathodic reaction (carbon electrode)

$$1/2\ O_2 + H_2O + 2e \longrightarrow 2\ OH^- \qquad (74)$$

Overall reaction

$$Zn + 2\ OH^- + 1/2\ O_2 \longrightarrow ZnO_2^{2-} + H_2O \qquad (75)$$

(2) Paramagnet method: Magnetic properties are not confined only to ferromagnetic materials. All substances are influenced by a magnetic field, although to a lesser degree than a material like iron. Substances attracted to a magnetic field are called "paramagnetic," and those repelled are called "diamagnetic." All liquids and gases fall into one or the other of these classes. Oxygen is unique among gases in that it possesses relatively high paramagnetic properties, whereas most gases are diamagnetic (Figure 62). Accordingly, it is possible to analyze for oxygen by means of its magnetic susceptibility.[2]

A carrier gas leaving the exchange unit passes through the oxygen-detecting unit containing a magnetic field in the vicinity of a hot spiral wire,[13] as shown in Figure 63. Upon being heated, molecular oxygen loses its magnetic properties and cooler oxygen molecules attracted into the magnetic field replace those that have been heated.

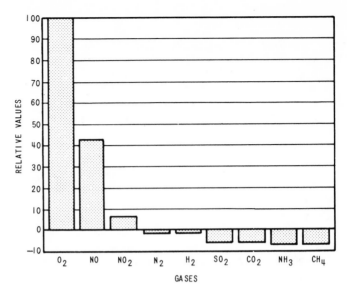

Figure 62. Magnetic properties of gases.

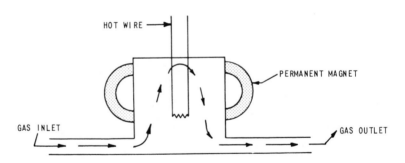

Figure 63. Paramagnetic analyzer cell.

This results in a circulation of gas in the vicinity of
the spiral wire. The change in the resistance of the
spiral wire caused by the flow of the cooler gas is proportional to the amount of oxygen present in the stream
of the carrier gas, which, in turn, is equal to the DO
content of the test solution. In another method for mag-
netically determining the oxygen content in the carrier
gas, a small glass dumbbell is suspended on a taut quartz
fiber in a nonuniform magnetic field (Figure 64). When

270

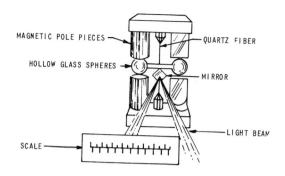

MAGNETIC POLE PIECES

QUARTZ FIBER

HOLLOW GLASS SPHERES

MIRROR

LIGHT BEAM

SCALE

Figure 64. Oxygen-sensitive paramagnetic detector.

oxygen is absent, the magnetic force exactly balances the torque of the quartz fiber and the dumbbell remains stationary. When the carrier gas containing oxygen is drawn into the detecting unit, the magnetic field is altered and the dumbbell rotates. The degree of rotation is proportional to the change in the magnetic force, which, in turn, is proportional to the oxygen concentration in the sample. By means of a small mirror attached to and rotating with the dumbbell, a beam of light is projected on a translucent scale, which is usually calibrated in terms of oxygen concentration.[3] (3) Thermal conductivity method: The thermal conductivity of a gas is defined as the quantity of heat transferred in unit time between two unit surfaces when they are a unit distance apart and the temperature difference between surfaces is 1°C.[8] The thermal conductivity of oxygen is usually determined in reference to the carrier gas (hydrogen or nitrogen).

Thermal-conductivity oxygen-detecting units contain a wire with a high temperature coefficient of resistance (*e.g.*, platinum) stretched along the axis of a metal or glass cylinder and connected to the arm of a Wheatstone bridge circuit, as shown in Figure 65. When a current is passed through the bridge, the wire becomes heated. The equilibrium temperature of the wire depends, however, on the thermal conductivity of the surrounding gas. Thus, changes in the thermal conductivity of the carrier gas due to the oxygen exchange cause an imbalance in the bridge. The imbalance is measured by a galvanometer or potentiometer, which can be calibrated in terms of the DO content of the test solution.[14]

Gas-exchange oxygen analyzers have been used for DO determinations in natural waters and wastes.[11,17,20,21,22]

271

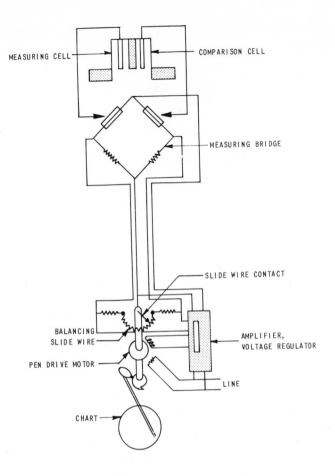

Figure 65. Schematic diagram of thermal conductivity oxygen analyzer.

The main advantages of these systems are:

1. Analyzing for oxygen in the gas phase makes it possible to eliminate the interference of ionic and organic compounds commonly present in natural water and waste.

2. The units are adaptable to continuous monitoring of DO.

3. These units are suitable for field operations; *e.g.*, they can be installed on the bank of a stream or on the side of an activated-sludge unit.

Gas-exchange oxygen analyzers, however, have the following drawbacks:

1. These oxygen analyzers are unable to determine DO *in situ* since they are not submersible.

2. The instruments are neither portable nor rugged to the extent that they can be mounted in a boat for a river survey.

3. These analyzers are not suitable for use on discrete samples in the laboratory.

The accuracy of the gas-exchange oxygen analyzers usually varies from ±2 to ±5 per cent of full-scale reading. Thus for a range of 0 to 5 mg/liter full-scale deflection, the accuracy is between ±0.1 to ±0.25 mg/liter. The precision of such instruments is about ±1 per cent of full-scale deflection.

Gas chromatographic methods

The method, as the name implies, depends on the determination of molecular oxygen in a gaseous stream.[42] The dissolved gases are stripped from the test solution with an inert gas. This is followed by chromatographic separation and subsequent detection of its components. The stripping is done in a gas exchange unit consisting of multiple spinning discs rotating at high speed, partially immersed in the water sample as shown in Figure 66. This unit is able to purge the water sample of its gas content in a single pass through, at inert gas/water flow ratios of 10 to 1 through 1 to 2.

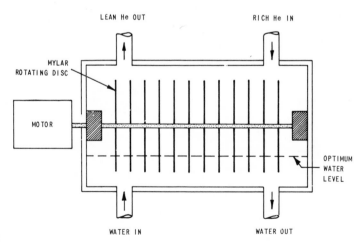

Figure 66. Gas exchange unit.

The gas stream from the exchange unit is then passed
through two gas chromatographic columns in series (Fisher
Gas Partioner, Model 25) as shown by the block diagram in
Figure 67. After the chromatographic separation of the
different components, the gas stream is passed through the
detector cells (thermal conductivity cells) and the elution curve is recorded by a 10-millivolt full-scale
recorder.

The instrument is calibrated in terms of peak height
versus gas volume, *e.g.*, millimeters peak height versus
millimeters of O_2 per liter. Good results were obtained
by using helium as the carrier gas. When argon was used,
the gas partitioner could not separate it from oxygen. A
suggested alternative procedure[42] involves the analysis of

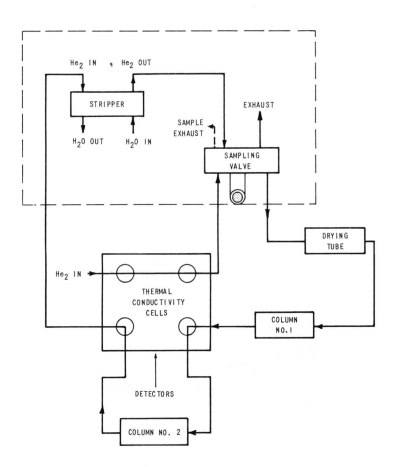

Figure 67. Gas chromatographic analyzer.

274

duplicate samples, one of which has had the oxygen removed by catalytic combustion or absorption before passing through the chromatographic columns.

The gas chromatographic method offers the following advantages:

1. Like the previously discussed gas exchange method, the chromatographic method eliminates the effect of interferences commonly present by analyzing for oxygen in the gaseous phase.

2. The chromatographic method is an on-stream device that can be adapted for continuous monitoring in plant operations.

3. This method can be used for the analysis for all gases in the test solution (CO, CH_4, N_2, and CO_2) simultaneously with oxygen.

The gas chromatographic method is, however, a relatively new method, and it has not been tried out in the field; therefore, an evaluation of its performance characteristics cannot be given. There are, however, certain inherent operational drawbacks, which may be summarized as follows:

1. It is a highly delicate, complex, instrument that requires specialized attention and constant regulation, *e.g.*, the regulation of carrier gas and the water flow rates.

2. The method is unable to analyze for DO *in situ*.

3. The instrument lacks the ruggedness and portability required for field use.

Radiometric method

A radiometric procedure for monitoring dissolved oxygen is based on the quantitative oxidation of radioactive thallium-204 by oxygen in the test solution.[30] Thallium-204 is primarily a beta emitter with a half-life of 3.6 years; therefore, decay over several months does not greatly reduce the sensitivity of the technique. The apparatus consists of a column of radioactive thallium electrodeposited on copper turnings, and two flow-type Geiger-Mueller counters (Figure 68).

The technique involves passing the test solution by one of the Geiger-Mueller counters to detect background beta-activity, then through the column where the following reaction occurs:

$$4Tl_{(s)} + O_2 + 2H_2O \rightarrow 4Tl^+_{(ag)} + 4OH^- \qquad (76)$$

The radioactive thallium in the effluent from the column is detected by the second Geiger-Mueller counter. One milligram of oxygen liberates 25.6×10^{-3} g of ^{204}Tl. The counting rate is directly proportional to the oxygen concentration in the test solution.

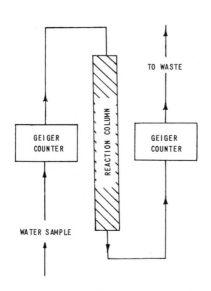

Figure 68. Radiometric method apparatus, employing thallium-204.

The sensitivity of the test using a column with a specific activity of 2.04 millicurie per gram of thallium is about 0.2 mg/liter. That is to say, a test solution containing 0.2 mg/l produces a ^{204}Tl counting rate equal to the background counting rate of the detector. As a rule, because of the randomness of radioactive disintegrations, the precision of this method is ±2%. It is important to note that oxidizing agents and changes in the pH of the test solution may interfere with the test.

Conductometric methods

Conductometric methods of analysis are based on the reaction of the DO of the test solution with compounds to form ionic species and increase proportionally the conductance of the test solution.

One method measures the conductivity of a flowing sample of water to which nitric oxide gas is added.[12] The DO oxidizes nitric oxide molecules to nitrite ions according to the following reaction.

$$4 \text{ NO} + O_2 + 2H_2O \longrightarrow 4H^+ + 4NO_2^- \quad (77)$$

The increased ionic strength of the test solution results in an increase in electrical conductance directly proportional to the DO concentration.

A flow diagram of this instrument is shown in Figure 69. The water sample passes first into ion-exchange columns to eliminate interferences from bases and buffer salts. Then it is passed through a reference conductivity cell and through the reaction column, where it comes in contact with the nitric oxide gas. The reaction of molecular oxygen with nitric oxide goes to completion instantaneously.[12] The sample is then passed through the second conductivity cell. The change in conductivity is usually recorded in terms of the DO concentration.

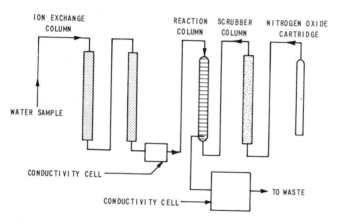

Figure 69. Conductivity apparatus.

In another approach, instead of using nitric acid gas, the water sample is passed into a column containing shavings of a solid reactant such as lead[44] or thallium metal.[45] The oxygen reacts with these metals to form their respective ions (anodic corrosion).

$$2 \text{ Pb} + O_2 + 2H_2O \longrightarrow 2Pb^{2+} + 4 \text{ OH}^- \quad (78)$$

or $$4 \text{ Tl} + O_2 + 2H_2O \longrightarrow 4 \text{ Tl}^+ + 4 \text{ OH}^- \quad (79)$$

The pH of the test solution is important since it influences the reactions.

Since the equivalent conductance of thallous hydroxide is 253, 1mg/liter of DO will produce an increase in

conductivity of 32 micromhos per centimeter (μmhos/cm).
It is important to note, however, that the background conductivity of different surface waters may vary from 100 to
1,000 μmhos/cm. To measure minute amounts of DO, the
water is first pretreated, if necessary, with ion-exchange
resins to insure a pH $\geq$ 7, reduce initial conductance of
the sample, and eliminate carbonates and phosphates, which
interfere with the measurements. After the water sample
has passed over the Pb or Tl surface, the change in conductivity is recorded. This may be expressed in DO concentration units.

Analysis by means of conductometric methods has been
used successfully in boiler feed waters. These methods
offer the following advantages:

1. There is a relatively high sensitivity of a few
   parts per billion.[45]

2. They are adaptable for continuous monitoring.

On the other hand, conductometric methods have the
following drawbacks:

1. They are incapable of analyzing for DO *in situ*.

2. Pretreatment of the sample by passing it through
   ion-exchange columns probably causes a loss of DO
   because of the large reaction surfaces. This, in
   turn, may activate reactions between molecular
   oxygen and organic or inorganic substances in
   the water sample.

Voltammetric methods

Voltammetric analyses for dissolved oxygen in waste
waters have been carried out with varying degrees of success using rotating platinum electrodes and dropping mercury electrodes. The main difficulty in using such
electrode systems in industrial waste effluents is the
presence of surface active and electroactive interferences
which frequently cause "electrode poisoning." A detailed
discussion of the effects of surface active agents on the
polarographic oxygen determination is available in the
literature.[25]

Various modifications of the dropping mercury electrode
system have been developed for continuous monitoring of
dissolved oxygen.[25,40] In the absence of interferences,
the sensitivity of this technique ranges from 0.05 to 0.10
mg of dissolved oxygen per liter.

Oxygen-sensitive galvanic cells have been used for some
time for analyses of water effluents.[25] These are made of
galvanic couples of an inert metal cathodes (*e.g.*, lead,
zinc or antimony).[39] The cathodic reduction of molecular
oxygen results in a galvanic current proportional to the
concentration of dissolved oxygen in the test solution.
Changes in the pH and the conductivity of the test solution influence the oxygen measurement.

Voltammetric membrane electrode

Perhaps the most dramatic development in the analysis
for DO since the Winkler test is the utilization of membrane electrode systems. These are particularly suitable
for applications in environmental science and engineering.
Only within the last 8 to 10 years, however, have membrane
electrode systems been used for DO determinations in natural and waste waters.

The basic structure of the oxygen membrane electrode is
composed of (a) a two solid electrode cell, (b) a thin
layer of supporting electrolyte in direct contact with the
electrodes and (c) an oxygen permeable membrane which
separates the electrodes and the electrolyte solution from
the test solution. To understand the principle of these
electrode systems, it may be helpful to consider conventional polarography.

In polarography the diffusion current is dependent
solely on the rate of diffusion of the electroactive
species in the test solution. The`test solution being of
no definite volume, the transport phenomenon is that of
infinite diffusion. In membrane electrode systems the
diffusion current is dependent solely on the rate of diffusion of the electroactive species in a finite layer,
which is achieved by placing a plastic membrane in close
contact with the electrode surface so that it separates the
electrode system from the test solution. Under these conditions the membrane constitutes a rigorously defined
diffusion layer, the thickness of which is independent of
the hydrodynamic properties of the test solution. Such
electrode systems exhibit more stable performance characteristics than the conventional types without a membrane
layer.

Oxygen membrane electrode systems are of two main types,
voltammetric membrane electrodes and galvanic membrane electrodes. The basic difference between voltammetric and
galvanic membrane electrode systems is that the former relies on an external source of applied voltage to polarize
the indicator electrode, similar to the conventional polarographic analysis. Galvanic analysis,[15] on the other hand,

is defined as a method that makes the species of interest a reactant at one of a pair of electrodes, thereby generating an electrical current without the aid of an electromotive force. Hence in the presence of the substance under galvanic analysis, a combination of electrodes (galvanic couple) and an electrolyte, the galvanic system becomes a primary battery or fuel cell in the general sense.

Several membrane electrode systems have been developed and used for the analysis for DO in natural waters.[5,7] For example the galvanic membrane electrode system[26] has been used widely and successfully in water pollution control programs and in waste treatment operations.[23,28] This electrode system is shown in Figure 70. It is composed of a silver-lead galvanic couple arranged in a ring-shaped anode surrounding a central disc-shaped silver cathode.[27,28] The galvanic couple is fitted to the tip of the plastic probe and is covered by a 0.5- or 1.0-mil-thick polyethlene membrane. A small amount of 4 M KOH, held by a small disc of lens paper, serves as a layer of electrolyte solution between the membrane and the electrode surface. After 2 to 4 weeks of operation the electrode system is recharged by cleaning the electrode surfaces, replenishing the supporting electrolyte, and mounting a new membrane.

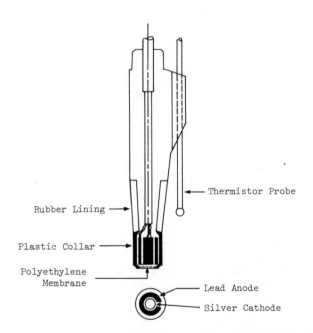

Figure 70. Galvanic cell oxygen analyzer.

With respect to circuitry, galvanic sensors are as simple to work with as thermocouples. All that is needed with the galvanic cell oxygen analyzer is a low-impedance galvanometer or microammeter. For recording, an inexpensive galvanometer-type recorder is adequate in most cases. Also, the cell circuit may be closed with a known resistor and the potential drop fed to a potentiometric recorder. Integrators, alarms, and controller equipment can be operated.

The load resistance in the circuit has little effect on the cell sensitivity since it is in series with the large polarization resistance of the sensing electrode. At high values of external load, however, the galvanic current is affected.

A galvanic membrane electrode that can be sterilized with steam has been used successfully for DO determinations in aerobic fermenters. This electrode system is made of a silver-lead galvanic couple and an acetate buffer as the supporting electrolyte.

Galvanic membrane electrodes, besides being self-energizing systems, have the advantage of being self-zeroing. In voltammetric membrane electrodes, an impressed voltage of either sign may give background signals.[15]

There are three main steps involved in the operation of oxygen membrane electrodes. The first is the permeation of molecular oxygen from the test solution through the plastic membrane layer. The second is the permeation of molecular oxygen through the electrolyte solution layer. The third step is the electrolytic discharge of oxygen on the silver cathode, with the subsequent generation of an equivalent quantity of current.

To aid in understanding the theory involved, the diffusion current equations as reported by the senior author[26] are presented briefly. Figure 71 shows a cross-sectional

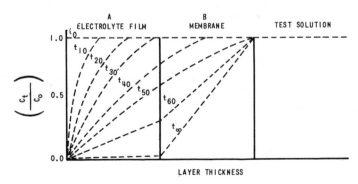

Figure 71. Concentration profiles.

diagram normal to the surface of the indicator electrode, which is composed of an electrolyte solution layer of thickness "A" and a membrane layer of thickness "B." Concentration profiles at different times of electrolysis, t, are also shown in Figure 71.

Figure 72 shows the corresponding current-time curve, in which steady-state operation was reached after 90 seconds. The current is assumed to be dependent solely on the rate of mass transfer (linear finite diffusion) of reactants to the electrode surface and not on the overall charge-transfer rate or on the kinetics of a chemical reaction coupled with the charge-transfer process. As soon as the electrolysis circuit is closed, the current shoots up to a high initial value, characteristic of charging the capacitance of the double layer, after which the current decreases in a manner determined by the rate of mass transport, as illustrated by the concentration profiles in

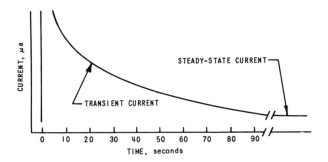

Figure 72. Current-time curve.

Figure 71. Current measurements at t=10 seconds, corrected for capacitance current thus depend on the concentration of the electroactive species in the electrolyte solution layer only. This can easily be seen upon consideration of the diffusion current equation for such two-film (the electrolyte film and the membrane) electrode systems. At very small values of t (*e.g.*, below 30 seconds in Figure 72), the electrolyte film is of primary importance and the current varies according to the following equation:

$$i_t = nFA\left(\frac{D_f}{\pi t}\right)^{0.5} a\left[1 + 2\sum_{n=0}^{\infty} e^{-\frac{n^2 L^2}{D_f t}}\right] \qquad (80)$$

where i_t is the current at time t, μa; n is the number of electrons exchanged per mole of reactant; F equals Faraday equals 96,600 coulombs; A is the surface area of indicator electrode, cm^2; D_f is the diffusion coefficient of the reactant in the electrolyte solution layer, cm^2/sec; a is strictly the activity of oxygen in test solution, M/cm^3; and A is the thickness of the electrolyte solution film, cm.

At very small t values, the summation term in Equation 80 can be neglected, and the current varies inversely with the square root of time.

$$i_t = \left[nFA \left(\frac{D_f}{\pi} \right)^{0.5} a \right] t^{-0.5} \qquad (81)$$

At larger values of $t (t \geq 30$ sec in Figure 71), diffussion in the membrane is governing.

$$i_t = nFA \frac{P_m}{b} a \left[1 + 2 \sum_{n=0}^{\infty} e^{-\frac{n^2 2 D_m t}{b^2}} \right] \qquad (82)$$

where P_m is the membrane permeability coefficient, cm^2/sec, b is the membrane thickness, cm, and D_m is the diffusion coefficient of reactant in membrane, cm^2/sec. At large values of t (above 90 sec), steady-state conditions are achieved and Equation 82 is then

$$i_{\infty} = nFA \frac{P_m}{b} a = nFA \frac{P_m}{b} \gamma C \qquad (83)$$

where γ is the activity coefficient and C is the concentration of dissolved oxygen.

The rate-determining step in this case is the transport through the membrane. Accordingly, the generated current is directly proportional to the rate of permeation of molecular oxygen through a given membrane, which, in turn, is directly proportional to the DO content of the test solution. A typical calibration curve showing the linearity of response under steady-state conditions is shown in Figure 73.

After the first day of continuous operation, the electrode reactions of the galvanic cell oxygen analyzer are partially as follows:

cathodic reaction

$$1/2 \ O_2 + H_2O + 2e \longrightarrow 2OH^- \qquad (84)$$

anodic reaction

$$Pb + 2 \ OH^- \longrightarrow PbO + H_2O + 2e \qquad (85)$$

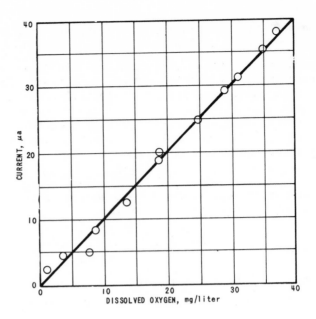

Figure 73. Calibration curve at 25°C.

overall reaction

$$Pb + 1/2\ O_2 \longrightarrow PbO \qquad (86)$$

It is evident then that as the cell discharges there is no net consumption of the OH^- concentration. Although the OH ions are not consumed in the overall cell reaction, they are required to sustain the anodic reaction. Hence, there will be a localized deficiency of OH^- ions at the anode and an increase in the OH^- ions in the electrolyte solution layer from the central cathodic area toward the peripheral anode.

Oxygen membrane electrodes have three main components: the membrane, the oxygen-sensing element, and the electrolyte solution. The membrane, the unique feature in such electrode systems, serves in three different capacities. First, the membrane acts as a protective diffusion barrier separating the sensing element from the test solution. Because plastic membranes are permeable to gases only, oxygen molecules pass through but electroactive and surface-active contaminants present do not. The possibility of poisoning the sensing element is thus eliminated. Second, the membrane serves to hold the supporting electrolyte in contact with the electrode system and thus makes it possible to determine oxygen in gaseous samples as well as in nonaqueous solutions such as industrial wastes.[26] The third advantage, discussed previously, is that the membrane constitutes a finite diffusion layer, the thickness of which is independent of the hydrodynamic properties of the test solution.

Membranes must meet the following requirements. Primarily, the membrane should have excellent permeability to oxygen, yet be practically impermeable to ionic species and water molecules from the test solution. Its permeability characteristics should not change with time, and there should be no intereaction between the membrane material and the diffusion oxygen molecules or ionic species present in the test solution. The membrane should be able to withstand slight amounts of stretching without breaking so that it can be applied easily to the electrode surface. Polyethylene, Teflon, polypropylene, and synthetic rubber show good permeability for oxygen. Polyethylene membranes of low crystallinity were found to have high permeability to oxygen.

As far as the oxygen-sensing element is concerned, platinum has been used extensively as cathode material in electrochemical analysis for DO. Silver and gold are, however, better cathode materials for oxygen detection than the conventionally used platinum. The reduction of oxygen on a platinum cathode was found to be a complicated process because the reaction proceeds through the formation of a platinum oxide layer. Significantly, this step does not occur with silver or gold electrodes.

The electrolyte solution in membrane electrode systems should be a highly conducting aqueous solution that does not interact with the electrode material. In voltammetric systems the type and concentration of the electrolyte solution are dictated by the reference anode. For example, for an Ag/AgCl reference electrode, a 3% KCl solution is used.

Sensitivity -- The membrane oxygen electrode is commonly calibrated by means of the Winkler test. A typical calibration curve is given in Figure 73, which shows a linear response to oxygen from zero to saturation values.

A basic operational requirement for membrane electrode systems is maintaining a threshold amount of mixing in the test solution around the tip of the analyzer. This can be achieved either by driving the solution past the surface of the analyzer by means of a pump or a stirring device, or by moving the analyzer. In the laboratory, a magnetic stirrer or a motor-driven glass stirrer is usually used.

285

Since the flow past the probe must be not less than lft/ sec, appropriate flow in the field may be attained by the natural flow of the water, by towing the probe behind a moving boat, or by moving the probe up and down manually. Attachments containing battery-driven small propellers[16] may be mounted on the tip of the probe.

The effect of stirring the test sample on instrument sensitivity is shown in Figure 74. From an initial low value with no stirring, the sensitivity rises rapidly with increased stirring speed and then assumes a steady value. Adequate stirring of the test solution is necessary so that the oxygen concentration at the membrane surface will be equal to that in the bulk of the solution. Only under such conditions will the measured diffusion current be directly proportional to the oxygen concentration in the bulk of the test sample.

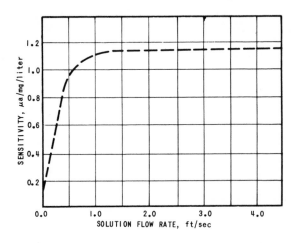

Figure 74. Effect of stirring on sensitivity.

Temperature Coefficient -- Another critical characteristic of voltammetric membrane electrodes is the relatively high temperature coefficient, which is largely due to the effect of temperature changes on the permeability characteristics of the plastic membrane. Temperature effects on the electrode reaction kinetics or the electrode potential are relatively negligible. Calibration curves for one of the galvanic cell oxygen analyzers over the temperature range from 5° to 35°C are shown in Figure 75.

The instrument sensitivity, ϕ, was found to vary with temperature (Figures 75 and 76).

$$\frac{d\phi}{dT} = m^{\circ} \frac{\phi}{T^2} \tag{87}$$

286

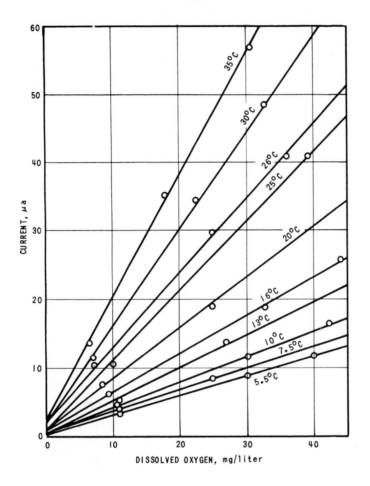

Figure 75. Calibration curves for different temperatures.

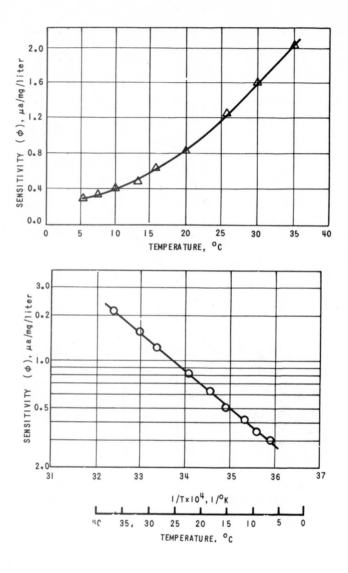

Figure 76. Effect of temperature on sensitivity.

where ϕ is the sensitivity, $\mu a/mg/liter$; T is the temperature, $^\circ K$; m° is the temperature coefficient, $^\circ K$.

If log ϕ is plotted versus $1/T$, a straight line is obtained.

$$\log \phi = \log \phi_0 + \frac{-m^\circ}{2.303} \left(\frac{1}{T_0} - \frac{1}{T}\right) \qquad (88)$$

in which $-m^\circ/2.303$ is the slope and b is the ordinate intercept. If m° is determined, the instrument sensitivity can be calculated for any desired temperature.

$$\log \phi = \log \phi_0 + \frac{m^\circ}{2.303} \left(\frac{1}{T_0} - \frac{1}{T}\right) \qquad (89)$$

Nomograph charts also can be constructed for the temperature correction.

Temperature compensation can be done automatically by incorporating a thermistor setting in the electrode circuit. Thermistors do not compensate fully for temperature errors to less than ±10% of the reading over the temperature range of 5° to $35^\circ C$, or less than ±5% over the 15° to $45^\circ C$ range.[2] When high accuracy is required, calibrated nomographic charts should be used.

Application -- Interpretation of the results of oxygen membrane electrode systems requires a great deal of care. It is helpful to point out the similarity between the glass membrane electrodes, used for the measurement of pH and pNa, and oxygen membrane electrodes. The measured parameter in the glass electrode system is an asymmetry potential across the glass membrane. This potential across the glass membrane is related to the hydrogen ion activity in the test solution.

In the case of oxygen membrane electrode systems, the measured diffusion current is solely dependent on the difference in the chemical potential of the electroactive species across the membrane. One should expect then that as glass electrodes are used for pH determinations (intensity), likewise oxygen membrane electrode systems will measure an "intensity property" that is essentially equivalent to the activity of molecular oxygen in solution. This should be differentiated from "extensive parameters" as determined by titrimetric methods such as the Winkler test. Accordingly, the difference between the pH value determined by the glass electrode and acidity as determined by titration with a standard base is essentially the same as the difference between oxygen analysis by a membrane electrode and the Winkler test. Only under ideal conditions and in the absence of certain impurities are diffusion current values from membrane electrode systems proportional to concentration. The correct form of the diffusion current equation for membrane electrodes is

$$i = \left(nFA \frac{P_m}{b}\right) a = \phi_a \, a = \phi_a \, \gamma C \qquad (90)$$

where ϕ_a is the sensitivity coefficient with respect to activity, $\mu a/M/cm^3$; γ is the activity coefficient of molecular oxygen; n is the number of electrons exchanged per mole of reactant; F is Faraday equals 96,500 coulombs; A is the surface area of the indicator electrode, cm^2; b is the thickness of the membrane layer, cm; P_m is the membrane permeability coefficient, cm^2/sec; a is the activity of the molecular oxygen, M/cm^3; and C is the concentration of molecular oxygen, M/cm^3.

The dependence of the diffusion current on the ionic strength of the test solution (Figure 77) can be expressed as follows:

$$i = (e^{K_s I} \, \phi_a) \, C \qquad (91)$$

where K_s is the salting-out coefficient and I is the ionic strength.

Equation 91 correlates the diffusion current, i, to the concentration of molecular oxygen in the test solution as a function of the ionic strength, I. In case I = 0, *e.g.*, distilled water, then the term $e^{K_s I}$ will be unity and Equation 91 will be equal to Equation 90. By means of

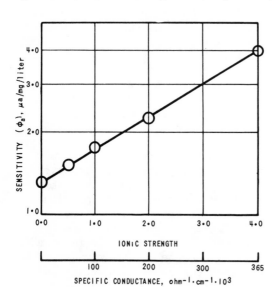

Figure 77. Effect of ionic strength on sensitivity.

290

Equations 90 and 91 values can be determined for the
activity coefficient, γ, and the salting-out coefficient,
K_s, in salt solutions (*e.g.*, KCl, $CaCl_2$, Na_2SO_4), which
are of high theoretical significance.

Recent investigations have shown that simple methods
can be developed to compensate for the effects of temperature and salt on the determination of DO concentration by
membrane electrodes.

The temperature effect can be described as follows:

$$i = \phi_{o,t}\, e^{-\frac{E_p}{R} \Delta \frac{1}{T}} C \qquad (92)$$

where $\phi_{o,t}$ is the sensitivity (current in ampheres/DO
concentration in M/liter) at a reference state with respect to temperature in $^\circ C$; E_p is the activation energy of
the permeation of the membrane, cal/M; T is the temperature, $^\circ K$; R is the gas constant, $cal/M/^\circ K$; C is the concentration of molecular oxygen, M/liter.

Similarly the effect of the salt content of the test
solution can be expressed as follows:

$$i = \phi_{o,s}\, e^{K_s \Delta I} C \qquad (93)$$

where $\phi_{o,s}$ is the sensitivity at a reference state with
respect to the ionic strength of the test solution amp/M/
liter.

The ionic strength of the test solution can be determined accurately for any given water by means of conductance measurements, as shown by the following equation:

$$L = \theta + (\lambda\, I) + (\delta\, I^2) + (\Gamma\, 'I^3) \qquad (94)$$

where L is the specific conductance, ohm^{-1}, cm^{-1}, and θ,
λ, δ, and Γ are constants.

The nonlinear relationship given in Equation 94 can be
approximated to

$$\Delta L = K_i\, \Delta I \qquad (95)$$

where K_i is the proportionality constant.

Equation 95 indicates the linear dependence of conductance on ionic strength, which holds in most surface
waters including estuarine water but does not hold in certain industrial wastes of high salt content.

By combining Equations 93 and 95, it follows that

$$i = \phi_{o,s}\, e^{K_s\, \Delta L}\, C \qquad\qquad (96)$$

where K_s equals $K_s K_i$.

The exponential terms in Equations 92 and 96 express the effect of the temperature and salt content.

Based on the above derivations, it is possible to develop an automatic compensator for changes in temperature and salinity. An example of such a compensator is shown in Figure 78.

Membrane electrode systems offer the following advantages:

1. The activity (and concentration) of molecular oxygen can be determined by using membrane electrode systems. Activity determinations are highly significant whenever environmental measurements and biochemical systems are concerned.

2. Since the sensing element is separated from the test solution by a selective permeable membrane, it is possible to analyze for DO in solutions containing ionic and organic contaminants.

3. The membrane serves as a diffusion layer, the thickness of which is independent of the hydrodynamic properties of the test solution. Accordingly, membrane electrodes exhibit more stable performance characteristics in flowing solutions than conventional electrode systems.

4. Membrane electrode systems possess the unique ability of analysis for DO in aqueous, nonaqueous, and gaseous systems.

5. One of the main advantages of membrane electrode systems is their suitability for field use. Because of their ruggedness, portability, and ease of operation and maintenance, they are ideal instruments for the analysis of DO under adverse environmental conditions.

6. Being completely submersible, they can be used to analyze for DO *in situ*.

7. Like other electrochemical systems, membrane electrodes can be used for continuous monitoring.

292

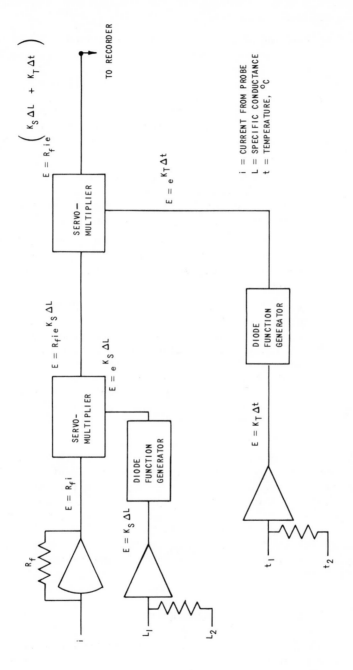

Figure 78. Compensator for temperature and salinity.

A DO determination by means of membrane electrodes is a relatively simple procedure; however, certain difficulties are sometimes encountered.

1. Most of the difficulties associated with these electrode systems are mainly caused by a lack of understanding of the procedure and principle of operation. In polarography, the electrode surface is continuously renewed, but in membrane electrode systems this is not the case. Great care should be taken during the recharging of the electrode system to avoid contaminating the sensing element.

2. Improper mounting of the membrane might lead to trapping small air bubbles under the membrane. The bubbles cause a slower response to oxygen and a higher residual current.

3. Insufficient flow across the membrane surface results in erratic response; maintaining a threshold mixing value in the test solution is therefore essential.

4. Plastic membranes show selective permeability to various gases and vapors. Gases reduced at the potential of the sensing electrode, *e.g.*, SO_2 and halogens, cause erroneous readings, but in aqueous systems these gases rarely exist in a free state. Other gases capable of permeating the plastic membrane may contaminate the sensing electrode or react with the supporting electrolyte, *e.g.*, CO_2 and H_2S.

Oxygen membrane electrodes have been used in a variety of water pollution control applications. Typical applications are oxygen monitoring in rivers and lakes, aeration studies, and oxygen utilization of domestic sewage.[25,27,28,40,41] Other applications include continuous DO recording in activated-sludge processes,[34,35] in respiration studies as a replacement of the Warburg Apparatus in oceanographic studies,[6,10,34] and in fermentation studies.[31,33]

REFERENCES

1. APHA, AWWA and WPCF. *Standard Methods - Water and Wastewater*, 12th ed. American Public Health Assoc. (1965).

2. Barbi, G., and S. Sandroni. Com. Nazl. Energia Nucleare CNI-55, 3 (1960).

3. Beckman Instrument, Inc. Continuous Oxygen Analyzers, Bulletin 0-4016, Beckman Instrument, Inc., Fullerton, California.

4. The Beckman Laboratory Oxygen Sensor. Bulletin 7013, Beckman Instruments, Inc., Fullerton, California.

5. Carritt, D.E., and J.W. Kanwisher. Anal. Chem. 31, 5, (1959).

6. Cassie, R.M. Calibration and Operation of Dissolved Oxygen Electrode, New Zealand Oceanographic Institute (1963).

7. Cleary, E.J. "An Electronic Monitor System for River Quality Surveillance and Research," Proc. 1st Intern. Conf. on Water Poll. Res., (New York: Pergamon Press, 1962).

8. Daynes, H.W. Gas Analysis by Measurement of Thermal Conductance (London: Cambridge University Press, 1933).

9. Dixon, W.S. "Pollution Control by Continuous Dissolved Oxygen Analysis and Associated Instruments," Proc. Water Quantity Measurement and Instrumentation, Technical Report, Robert A. Taft Sanitary Engineering Center, Cincinnati, Ohio (1960).

10. Duxbury, A.C. Limnol. Oceanog. 8, 483 (1963).

11. Dykes, D.F., and J.C. Baumgartner. "Use of Pure Oxygen for Aerating Craft Mill Wastes," Proc. 6th Southern Munic. and Ind. Wastes Conf., N.C. State Coll. (1957) p. 200.

12. Finnegan, T., and R. Tucker. "The Beckman Dissolved Oxygen Analyzer," ASTM Special Technical Publication No. 219 (1959).

13. Hays Thermal Conductivity Analyzer. The Hays Corporation, Michigan City, Ind. (1960).

14. Hays Magno-Therm Oxygen Recorder. The Hays Corporation, Michigan City, Ind. (1960).

15. Hersh, P.A. "Galvanic Analysis," *Advances in Analytical Chemistry and Instrumentation*, Vol. III, C.N. Reilly, Ed. (New York: Academic Press, 1964) p 210.

16. Hissel, J., and J. Price. Bull. Centre Belge Etude et Document Eaux, 44, 76 (1959).

17. Interstate Commission on Potomac River Basin "News Letter," 18, 7 (1962).

18. Keidel, F.H. Ind. Eng. Chem. 52, 490 (1960).

19. Lederer, V.V. Chem. Listy 41, 230 (1947).

20. Levin, H.S., and R.S. Kleinschmidt. Sewage Ind. Wastes 29, 856 (1957).

21. Levine, H.S., *et al.* Anal. Chem. 28, 343 (1956).

22. Macklin, M.O., D.J. Baumgartner, and M.B. Ettinger. Sewage Ind. Waste **31**, 804 (1959).

23. Mancy, K.H., and J.E. Haskins. "The Voltammetric Analysis of Dissolved Oxygen. I. The Salt Effect," 148th ACS Meeting, Chicago (Sept. 1964).

24. Mancy, K.H., and T. Jaffe. Environmental Health Series, Public Health Service Publication **999-WP-37** (1966).

25. Mancy, K.H., and D.A. Okun. Anal. Chem. **32**, 108 (1960).

26. Mancy, K.H., and D.A. Okun. "Oxygen in Waste Treatment - the Galvanic Cell Oxygen Analyzer." Special report to NIH, Dept. of Environmental Sciences and Engineering, School of Public Health, University of North Carolina, Chapel Hill, N.C. (1961).

27. Mancy, K.H., D.A. Okun, and C.N. Reilly. J. Electroanal. Chem. **4**, 65 (1962).

28. Mancy, K.H., and W.C. Westgarth. JWPCF **34**, 1037 (1962).

29. Morris, J.C. Final report, Contract SAph 69705, Department of Health, Education, and Welfare with Harvard University, May 1, 1958 to January 31, 1959.

30. Potter, E.C., and J.F. Moresky. "The Use of Cation-Exchange Resin in the Determination of Copper and Iron Dissolved at Very Low Concentrations in Water," in *Ion Exchange and Its Applications* (London: Society of Chemical Industry, 1955) p 93.

31. Rideal, S., and G. Stewart. Analyst **26**, 141 (1901).

32. "Solubility of Atmospheric Oxygen." Twenty-ninth progress report of the Committee on Sanitary Engineering Research of the Sanitary Engineering Division of ASCE, Proc. ASCE **86**, 41 (1960).

33. Strohm, J.D., and R.F. Dale. Ind. Eng. Chem. **53**, 760 (1961).

34. Sullivan, J.P. "Determination of Dissolved Oxygen and Nitrogen in Sea Water by Gas Chromatography," *Marine Science Department, Report No. 0-17-63* (Washington, D.C.: U.S. Naval Oceanographic Office, 1963).

35. Sulzer, F., and W.M. Westgarth. Water and Sewage Works **8**, 266 (1962).

36. Swinnerton, J.W., V.J. Linnenbom, and C.H. Cheek. Anal. Chem. **34**, 483 (1962).

37. Szekely, G., G. Raoz, and G. Traply. Period. Polytech., Chem. Engr., **10**, 231 (1966); Chem. Abstracts, **67**, 500 (1967).

38. Thayer, L.C., and D.A. Robinson. Proc. First International Congress of Instrument Society of America, Philadelphia, **9**, 2 (1954).
39. Todt, F., and R. Freier. Ger. Pat. 1-012-764 (1957).
40. Tyler, C.P., and J.H. Karchmer. Anal. Chem. 31, 499 (1959).
41. Weiss, C.M., and R.T. Oglesby. Instrumentation for Monitoring Water Quality in Reservoirs. Presented at 83rd Annual Conference of AWWA, May 23, 1963.
42. Williams, D.D., and R.R. Miler. Anal. Chem. 34, 657 (1962).
43. Wilson, R.F. Limnol. Oceanog. 6, 259 (1961).
44. Wilson, Ray I., and Jan N. Haagen-Smith. "Measurement of Dissolved Oxygen," U.S. Pat. 3,042,495 (July 3, 1962).
45. Wright, J.M., and W.T. Lindsay, Jr. "New Method for the Continuous Analysis of Dissolved Oxygen in Water," Proceedings of Am. Power Conf. 21, 706 (1959).

The basis of any water pollution control program is data. Most of it will have to be provided by chemists, some of it by engineers. It is, therefore, essential that an analytical chemist be familiar not only with chemical analysis but also with the collection, analysis and interpretation of data.

Typical data collection might be:
  (1) the calibration of analytical instruments,
  (2) an analysis of the precision of instruments,
  (3) complete chemical analysis of water and waste water,
  (4) the collection of records on the performance of water and waste water treatment plants, and
  (5) the monitoring and surveillance of water quality in river basins.

Associated with any data collection effort are a number of rather difficult questions, some of which might be:
Why should the data be collected?
Specifically what data are needed?
Where should it be collected?
At what time and intervals should it be obtained?
How many are necessary?
What will be done with the data?
The answers to these questions vary, depending on each situation.

To begin with, statistical analysis is performed in the following sequence:
  (1) collection of the numerical information,
  (2) classification of the data based on the information sought, and

(3) generalization of the classified information.
The latter step means that statistical measures are sought
which summarize the data in convenient form. For example,
the average of different samples is a meaningful statistic,
the variation serving as an indicator of the variability
of the samples. Or, it might be interesting to know if
time-dependent data exhibit a certain trend, or fluctuate
in a cyclical form. There is one caveat in this generalization: it is always assumed that the limited data
gathered are a representative sample of the "total population" or "universe" and that other like samples would
show the same relationships.

In a short chapter like this, it is impossible to men-
tion all the statistical techniques, and the reader is
referred to the standard statistical textbooks, some of
which are listed at the end of this chapter. Only some of
the more frequently needed techniques such as curve
fitting and tests for statistical significance are
discussed.

Curve Fitting

The first and probably most difficult step in curve
fitting is the selection of a suitable equation. In some
cases there is a rational basis for selection of an equation, because the underlying physical and chemical principle dictates a certain equation. However, quite often,
the approach is quite empirical.

On graphing the data, it is wise to observe a number of
rules.

(1) The independent variable should be plotted
    along the x-axis (abscissa). The dependent
    variable should be plotted along the y-axis
    (ordinate).

(2) The scales should be chosen in such a way
    that they are easily recognizable and that
    the graph covers the entire paper.

(3) The variables should be chosen in such a
    way as to result in a straight line graph.

Plotting data on ordinary graph paper will usually indicate either a curve or straight line. If the data exhibit a straight line relationship, the line of best fit
may be drawn in by eye thus determining the coefficients
of the equation. On the other hand, if the data follow a
curve, it will be necessary to fit a curve to the data.
Because the human eye is not very good at doing this, it

is preferable to use a transformation of the x and/or y data to obtain a straight line relationship. The transformation necessary to change selected curves into straight lines is shown in Figure 79.

Another possibility is the use of special functional paper, such as semilog or log-log paper. This aids greatly, since no transformation of the original data is necessary.

Example 1

To study the influence of mixing on a given cell-electrode system in anodic stripping voltammetry, a series of experiments were performed in which the response of the system was determined as a function of the rpm of the stirring blade (Table 35). In this kind of experiment one expects that increasing the rpm beyond a certain limit will not lead to an increase in the current response,

Table 35

rpm	*Current* μA
0	1.5
50	3.9
100	5.0
150	7.9
200	9.5
250	11.1
300	12.2
350	13.8
400	14.8

since this response is a function of thickness of the diffusion layer on the electrode which approaches a stable minimum after a certain rpm is reached. On the other hand, when there is no mixing, there is still a response of the electrode because of diffusion only.

Plotting these data on regular arithmetic paper led to a curve similar to the sixth one of Figure 79, except that it would not go through the origin. It was therefore postulated that the equation would be of the form

$$y = c + \frac{x}{a + bx}$$

If the data follow these curves,	Use this equation, and	Plot data using transformed equation
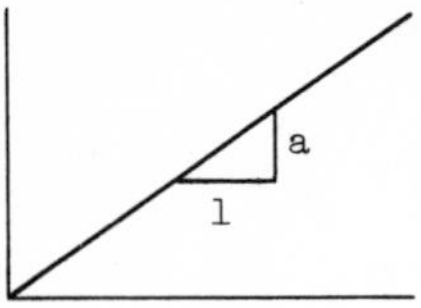	$y = ax$	$y = ax$
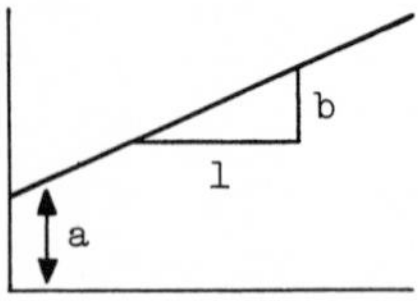	$y = a + bx$	$y = a + bx$
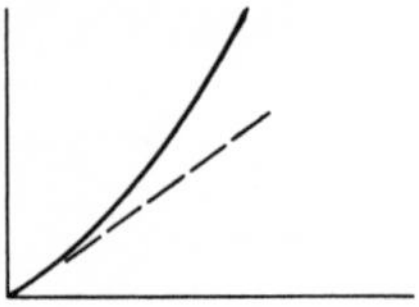	$y = ax^b$	$\log (y) = \log (a) + b \log (x)$
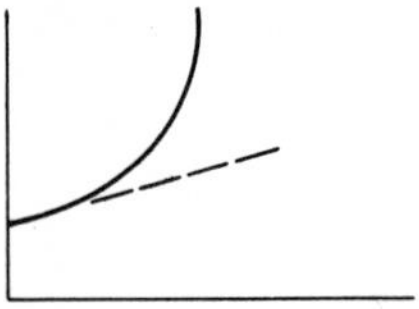	$y = ae^{bx}$	$\log (y) = \log (a) + (b \log e) x$
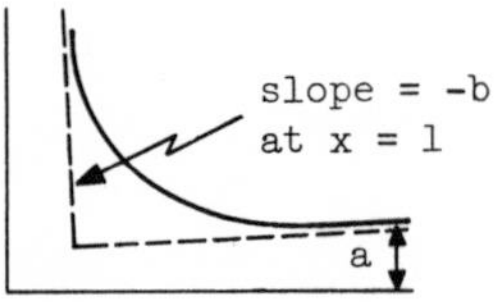	$y = a + \dfrac{b}{x}$	$y = a + b(\dfrac{1}{x})$
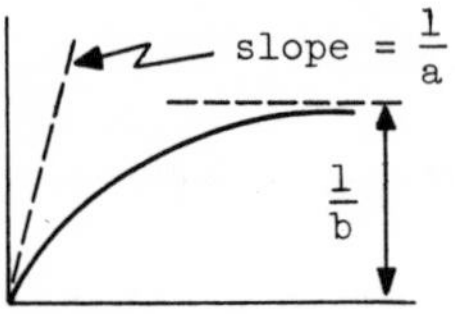	$y = \dfrac{x}{a + bx}$	$(\dfrac{x}{y}) = a + bx$
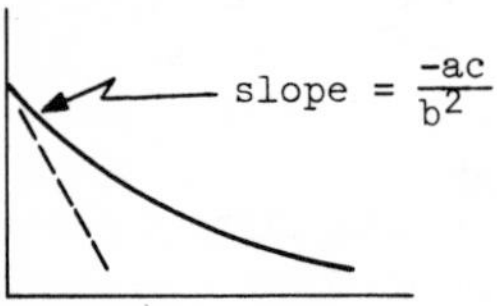	$y = \dfrac{a}{b + cx}$	$(\dfrac{1}{y}) = (\dfrac{b}{a}) + (\dfrac{c}{a})x$

Figure 79. Most Frequently Encountered Curves, Their Equations, and the Transforms to Linearize the Equations.

where y is the current and x the rpm of the stirring blade. After subtracting C = 1.5 from all data, the ratios of x/y were plotted against x. This resulted in the graph of Figure 80. A visual fit through the data points led to a straight line passing through the points (20,0) and (30,400). The constants of the line are a = 20, and b = (30.-20)/400. = .025. And thus the equation describing the current as a function of the rpm's of the blade is

$$y = 1.5 + \frac{x}{20 + .025x}$$

In a number of cases the researcher does not wish to rely on his eyes only but would like to use arithmetic procedures to calculate directly the parameters of his selected equation based on the data. This is usually done by the so-called "method of least squares," which is designed to minimize the sum of the squares of the differences between observed and calculated values in either x or y direction. The usual procedure is to minimize the sum of the squared differences in y direction.

Consider, for example, the case where an equation of the form y = a + bx is to be fitted to n pairs of observations x and y. Let the difference for the i-th pair between the observed value y_i and the calculated value $(a + bx_i)$ be denoted by d_i, then

$$d_i = (a + bx_i) - y_i \qquad (97)$$

The sum of the squares of the differences then is

$$\sum_{i=1}^{n} d_i^2 = \sum_{i=1}^{n} |(a+bx_i) - y_i|^2 \qquad (98)$$

The two unknowns to be determined are a and b. Taking the partial derivatives with respect to a and b leads to the necessary conditions for a minimum, namely

$$\frac{\partial \sum d_i^2}{\partial a} = 2\sum (a + bx_i - y_i) = 0 \qquad (99)$$

$$\frac{\partial \sum d_i^2}{\partial a} = 2 \sum (a + bx_i - y_i) x_i = 0 \qquad (100)$$

Rearranging both equations leads to:

$$a + b(\frac{\sum x}{n}) - (\frac{\sum y}{n}) = 0 \qquad (101)$$

$$a (\frac{\sum x}{n}) + b(\frac{\sum x^2}{n}) - (\frac{\sum xy}{n}) = 0 \qquad (102)$$

303

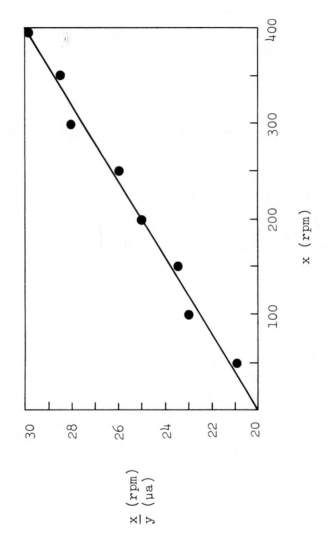

Figure 80. Plotting of the ratio rpm/current versus rpm.

In the special case when it is postulated that the curve
must pass through the origin, that is, a = 0, one obtains

$$b = \frac{\Sigma xy}{\Sigma x^2} \qquad (103)$$

Example 2

In anodic stripping voltammetry it is necessary to calibrate a specific cell-electrode system. For example, to
determine the sensitivity of such a system to cadmium
ions, a series of experiments was performed with a varying,
but known, cadmium ion concentration. The peak current
was measured for each cadmium concentration. Table 36
gives cadmium concentrations versus the peak current.
Based on the equations governing anodic stripping voltammetry, the relation between concentration and peak current
must be linear and pass through the origin. Find the calibration curve.

Table 36

Cadmium Conc. $10^{-9}mol/1$	Peak Current µA
25	2.7
50	6.3
75	10.3
100	13.7
125	17.6
150	22.5

In this example we have to estimate only one coefficient, namely the slope b. The necessary calculations are
summarized in Table 37, where b = 8,099./56,875. = 0.142.
Therefore, the sought equation is

Current (µA) = .142 x Cadmium Conc. (10^{-9} mol/1)

305

Table 37

x	y	x^2	xy
25.	2.7	625.	67.5
50.	6.3	2,500.	315.0
75.	10.3	5,625.	772.5
100.	13.7	10,000.	1,370.0
125.	17.6	15,625.	2,200.0
150.	22.5	22,500.	3,375.0
Sums 425.	73.1	56,875	8,099.0

Example 3

Consider now the problem of fitting a straight line to the following set of observations:

y	3	3	7	6	12
x	1	2	4	5	8

We shall assume that the curve to be filled is of the form

$$y = a + bx$$

Based on the equations developed before, the necessary computations are given in Table 38.

Table 38

Observed y	Observed x	x^2	$x \cdot y$	Calculated y
3	1	1	3	2.3
3	2	4	6	3.6
7	4	16	28	6.2
6	5	25	30	7.5
12	8	64	96	11.4
Sums 31.	20.	110.	163.	31.0
Means 6.2	4.0	22.0	32.6	6.2

Using the equations developed before, one obtains:

$$a + 4b = 6.2$$

$$4a + 22b = 32.6$$

Multiplying the first equation by 4 and subtracting it from the second equation leads to

$$6b = 7.8$$

$$b = 1.3$$

Inserting the value of b into the first equation yields

$$a = 1.0$$

Therefore, the least squares fit is the equation:

$$y = 1.0 + 1.3x$$

This expression may be thought of as "mean regression" because the minimum of a sum of squares is taken on at the mean.

Alternatively, one might wish to minimize the sum of the absolute deviations and call this "median regression."

$$\min_{a,b} \sum_{j=1}^{5} \left| y_j - a - bx_j \right|$$

When casting this problem in a linear programming* form, we note we can represent any variable as the difference of two non-negative variables. We will therefore add a non-negative slack variable u and subtract a non-negative slack variable v from each observation to account for the ± deviations.

Since we wish to minimize the total sum of the absolute deviations, we have the following linear programming problem:

Minimize: $\quad u_1 + v_1 + u_2 + v_2 + u_3 + v_3 + u_4 + v_4 + u_5 + v_5$

Subject
to:
$$3 = a + b + u_1 - v_1$$
$$3 = a + 2b + u_2 - v_2$$
$$7 = a + 4b + u_3 - v_3$$
$$6 = a + 5b + u_4 - v_4$$
$$12 = a + 8b + u_5 - v_5$$

The solution of the linear programming problem is $a = 1.71$, $b = 1.29$. We observe that these two "optimal" fits are

---

*For the theory and development of linear programming, the reader is referred to the standard textbooks on this subject.

different. Figure 81 shows the differences. Also note
that the minimum absolute deviation (M.A.D.) fit passes
through the two points (1,3) and (8,12). In general, if
the estimates of a and b are not zero, the M.A.D. fit will
pass through at least two points.

There is a third possibility of fitting a straight line
through the data, namely, in such a way as to minimize the
maximum absolute deviation. If one defines the maximum
absolute deviation as e, then the problem may be formulated
as

$$\text{minimize } e$$

subject to:

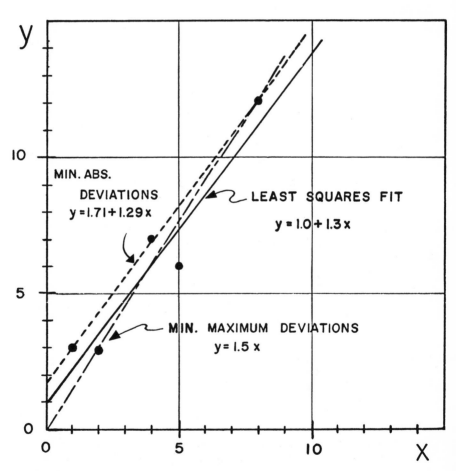

Figure 81. Using Different Criteria for Fit Leads to
Different Equations.

$$a + bx_i - e \leq y_i$$

$$i = 1, 2, \ldots, 5$$

$$a = bx_i - e \geq y_i$$

Again this is a linear programming problem which may be solved using the standard logarithms. The solution is y = 1.5 x. All three fits are shown in Figure 81.

Use of Probability Paper

If we have a large number of observations from some universe, it is an interesting fact that a plot of the relative frequency against the magnitude will produce a symmetrical, bell-shaped distribution known as the gaussian or normal curve. The mathematical explanation given for this fact is:

(1) the largest number of observations should occupy the center,

(2) values greater than the central magnitude occur as often as values that are smaller, and

(3) smaller and smaller numbers of observations should have a larger and larger distance from the central magnitude.

The shape of this curve is completely defined by two statistical parameters. First, the mean or average is defined as

$$\bar{x} = \sum_{i=1}^{n} x_i$$

where n is the number observations and x_i is the i-th observation. Secondly, the standard deviation is a measure of the width or spread of the curve, and is defined as

$$\sigma = \frac{\sum (x_i - \bar{x})^2}{n - 1}$$

The well-known properties of the normal distribution are that roughly 68 per cent of the data lie between $\bar{x} \pm 1\sigma$, 95.45 per cent of the data lie between $\bar{x} \pm 2\sigma$, and 99.7 per cent lie between $\bar{x} \pm 3\sigma$. These limits do not exactly apply for small samples but strictly apply only for large n.

While the two parameters of the normal distribution can be calculated directly from the two equations described before, it is often desirable to use graphical

309

methods to determine them. For this purpose special graph
paper has been developed which transforms the cumulative
probability curve into a straight line. Thus data which
follow a normal distribution should plot as a straight
line on this paper. To determine the use of this paper,
consider the following example.

Example 4

The efficiency of an industrial waste treatment plant
is to be analyzed. There are 19 samples available, which
are listed in Table 39. Determine the mean efficiency and
the variation in the data using normal probability paper.

Table 39

Efficiency	Efficiency in Order of Magnitude	Serial Number (m)	Plotting Position m/(n+1)
90	76	1	5.0
81	79	2	10.0
84	81	3	15.0
86	81	4	20.0
83	81	5	25.0
76	83	6	30.0
85	84	7	35.0
79	84	8	40.0
87	84	9	45.0
92	85	10	50.0
84	85	11	55.0
93	86	12	60.0
81	87	13	65.0
88	87	14	70.0
90	88	15	75.0
81	90	16	80.0
85	90	17	85.0
87	92	18	90.0
84	93	19	95.0

The first step in such an analysis is to organize the
data in increasing order. This was done and is shown in
column 2 of Table 39. Next, a serial number is assigned
as shown in column 3 of Table 39. Then, a plotting position is computed, using the equation m/(n+1) * 100 as
shown in column 4 of Table 39. Finally, each point is

310

plotted using the plotting position as the x-coordinate and
the efficiency as y-coordinate. If the data points describe a straight line, a line is drawn through these points
by eye. This was done as shown in Figure 82. The major
statistical parameters, namely mean and standard deviation,
may be read directly from the graph. The mean is determined by finding on the probability scale (x-axis) the
value of 50, extending this line vertically until it in-
tersects the fitted line, and proceeding then from this
point to the vertical scale (y-axis). In the example, the
mean of the data is shown to be 85. To determine the
standard deviation, recall that 68 per cent of the data
will fall within one standard deviation from the mean.
Proceeding therefore from the intersection of (50 - 34) =
16 per cent and (50 + 34) = 84 per cent probability, one
finds the intersection with the fitted line and determines
then the corresponding intersection on the y-axis. In the
example the two intersections on the y-axis are 80 and 90.
The difference is equal to 10 and corresponds to 2σ.
Therefore, the value of σ is determined graphically to be
5. For greater accuracy in these calculations it is usual
to use the data corresponding to $\pm 2\sigma$ since this gives a
better estimate of σ. In this case we would obtain (95 -
75)/4 = 5 as before.

Several points are worth noting. First, if the data
do not fall in a straight line, they do not follow a normal distribution and another distribution should be tried.
Second, individual data points which are "out of line" are
immediately spotted. In some cases this is an indication
that something has gone wrong and that this experiment
should be performed again or that this data point should
be disregarded. Third, it should be noted that the slope
of the line is a measure of the standard deviation; the
steeper the slope, the larger the standard deviation. And
fourth, other statistics can be read directly off the
graph. For example, if one is interested in determining
the probability that one of the samples is less than, say
74, simply find the intersection with a horizontal line at
74 and the corresponding vertical probability line. In
the example, the probability for this would be about one
per cent. Thus there is less than a one-in-a-hundred
chance that we would obtain a sample of less than 74.

If the data plotted on arithmetic probability paper do
not form a straight line but exhibit a concave curve, it
may be that the logarithms of the data are normally distributed. Many measurements of biological phenomena, such as
the MDN of coliform organisms, follow such a distribution.
Furthermore, this distribution has a natural lower limit

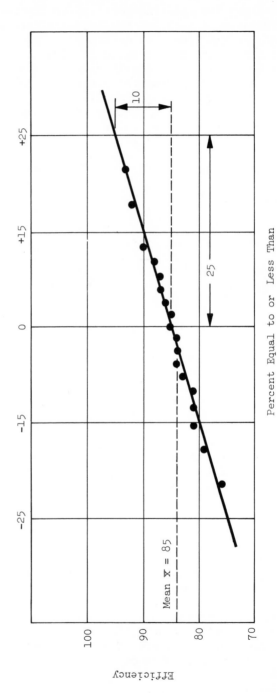

Figure 82. Plot of the Efficiency Data of Table 39 on Normal Probability Paper.

of zero, and thus all phenomena which have such a lower limit probably follow better the log-normal distribution. However, in many cases it is not possible to anticipate what distribution the data will follow and the only recourse is to try both distributions. Again, as for the normal distribution there is graph paper available which will transform the cumulative log-normal distribution into a straight line. This is shown in the following example.

Example 5.

The effluent BOD of an industrial water treatment plant is shown in Table 40. Find mean and standard deviation.

Table 40

Observed BOD	Ordered BOD	Serial Number	Plotting Position
12	3	1	5
15	4	2	10
4	4	3	15
7	5	4	20
9	6	5	25
9	6	6	30
20	7	7	35
30	7	8	40
6	9	9	45
7	9	10	50
6	9	11	55
3	12	12	60
4	12	13	65
5	12	14	70
9	15	15	75
12	15	16	80
12	20	17	85
15	20	18	90
20	30	19	95

Columns 2, 3 and 4 of Table 40 are prepared just as before, namely, ordering the data in increasing order of magnitude, assigning a serial number to them, and calculating a plotting position based on the formula m/(n+1)*100, where m is the serial number and n is the number of data points. The plot of the data is shown in Figure 83. The points follow a straight line fairly well; their mean is a BOD of

313

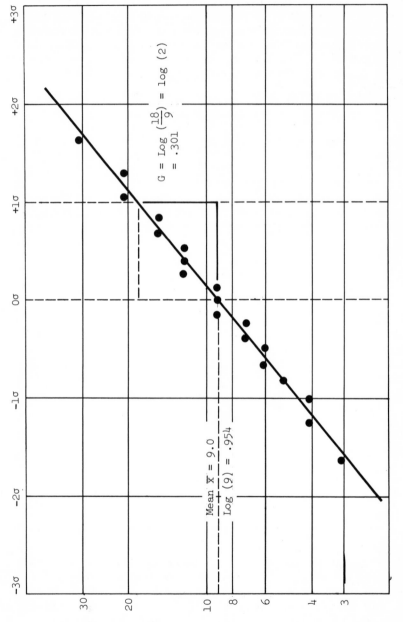

Figure 83. Plot of the BOD Data of Table 40 on Logarithmic Probability Paper.

9.0, and the standard deviation is 2. This means that roughly 68 per cent of the data lie between 9.0/2.0 = 4.5 and 9.0 x 2.0 = 18.

There are other probability papers available which will allow the straight line plotting of specific probability distributions, such as the Gumbel's extreme value distribution. But these are rather special applications.

Tests for Statistical Significance

Quite frequently questions of the following nature arise: is method A as good as method B, or is there a significant difference? One of the standard statistical tests to answer such a question is the so-called t-test.

The t-test is used to test the hypothesis that two means do not significantly differ. The test is useful for the comparison of results obtained by two different methods of analysis or by two analysts. Many modifications of the test are possible, including those for paired and non-paired observations and whether the variances are the same or different. To determine if the variances, V, of the two populations are the same, the variance ratio, F, is computed by the formula

$$F = \frac{s_1^2}{s_2^2} = \frac{V_1}{V_2}$$

where $V_1 > V_2$. The calculated ratio is compared with the ratio from standard statistical tables. If the calculated ratio is less than the tabulated value, the variances are not statistically different and may be pooled for use in the t-test by the equation

$$V = s^2 = \frac{\Sigma(x_1 - \overline{x}_1)^2 + \Sigma(x_2 - \overline{x}_2)^2}{n_1 + n_2 - 2}$$

where $\overline{x}_i$ is the mean and n_i is the number of observations in the i-th sample. The value of t is

$$t = \frac{\overline{x}_1 - \overline{x}_2}{s} \left(\frac{n_1 n_2}{n_1 + n_2}\right)^{1/2}$$

This value is compared to tabulated values for $n_1 + n_2 - 2$ degress of freedom. Tables will give t values at various confidence levels (e.g., 90, 95, 99, 99.9%). The analyst must be cautious in choosing the confidence level. If too high a level is selected, significant differences may not be detected. If the level selected is too low, an insignificant difference may be judged significant.

Table 41 presents the data and calculations for the comparison of two methods of analysis applied to the same sample. Six analyses were made by method 1 and seven by method 2. The averages were 11.9 and 13.1, respectively.

Table 41

Comparison of Two Methods of Analysis By Statistical Testing

Method 1: 11.3, 12.2, 11.5, 13.0, 10.5, 12.4

Method 2: 13.5, 13.0, 14.0, 12.1, 11.9, 14.5, 12.7

Averages and Variances:

$$\bar{x}_1 = 11.9 \qquad v_1 = 0.806$$
$$\bar{x}_2 = 13.1 \qquad v_2 = 0.923$$

F test:

$$F = \frac{0.923}{0.806} = 1.15$$

Variance based on both methods:

$$v = \frac{4.03 + 5.54}{11} = 0.870$$

$$s = (0.87)^{1/2} = 0.933$$

t-test:

$$t = \frac{11.9 - 13.1}{0.933} \left(\frac{6 \times 7}{13}\right)^{1/2} = -2.31$$

Before computing the t value, an F ratio must be computed to test for homogeneity of variances. The calculated value 1.15 is compared with the tabular value for six degrees of freedom for the numerator and five for the denominator. The calculated value is less than 4.95, which is the tabular value at the 95% confidence level. The two variances are not statistically different and a variance based on both methods can be calculated. The variance based on both methods is 0.870; the standard deviation is 0.9333. The calculated value of t, disregarding the sign, is 2.31, which should be compared to the tabular values of 2.201 for 95% and 3.106 for 99% confidence levels.

Since the calculated value of t is greater than 2.201, in at least 95 cases out of 100 such a t value indicates a true difference between the two methods while 5 times out of 100 this value could have occurred by change alone. Since the calculated t value is less than 3.106 it cannot be stated with 99% certainty that the methods give different results. A result which is positive at the 95% level is regarded as significant, while a positive result at the 99% level is highly significant. In this case, it appears that the two methods give different results, but to decide with greater certainty, more tests should be run.

Another possible use of the t-test is to determine if the correlation coefficient of two series of measurements is significant. The correlation coefficient is a popular index of the degree of association between two continuous variables and varies between -1 and +1. If the correlation coefficient is zero, we have no correlation; if the coefficient is 1, we have complete and perfect correlation and we know that there must be some fixed relation between the two variables. Thus, for example, if we find a high degree of correlation between two parameters of water pollution, we may measure only one and determine the other from a mathematical relation to be determined. The correlation coefficient between a series of measurements x_i and y_i is defined as

$$ r = \frac{\Sigma(x_i - \bar{x})\,(y_i - \bar{y})}{\sqrt{\Sigma(x_i - \bar{x})^2\,\Sigma(y_i - \bar{y})^2}} $$

If we are interested in testing whether the calculated value is different from zero, that is, whether the hypothesis of independence is contradicted by the data, we first calculate r and then determine the quantity t by

$$ t = \frac{r\sqrt{n-2}}{\sqrt{1-r^2}} $$

which is exactly a "student's" t with $n - 2$ degrees of freedom. We can, therefore, use the standard tables which give the values of r that must be exceeded for significance at various levels according to the degree of freedom available. If we had a total of 16 pairs of observations (14 degrees of freedom), then only a value of r greater than .497 or less than -.497 would be significant at the 5 per cent level.

Reliability of Statistical Results

In interpreting the results of a statistical analysis, the reliability of these results becomes important. Generally speaking, the larger the number of samples, the better the reliability. For example, to find the degree of reliability of the mean, it is necessary to investigate how a number of means of different subsamples are distributed. These statistics are well developed for the arithmetically normal probability distribution and are listed below:

Parameter		Standard Deviation of Parameter
Mean	$M = \Sigma x_i / n$	$\sigma_M = \sigma / \sqrt{n}$
Standard Deviation	$\sigma = \sqrt{\Sigma(x_i - M)^2 / (n - 1)}$	$\sigma_\sigma = \sigma \sqrt{2n}$

This information may then be used to answer the following type of questions: Given a sample mean and a standard deviation, what are the confidence intervals for the true population mean? In Example 4 the mean efficiency of the industrial waste treatment plant was found to be 85 with a standard deviation of 5. The standard deviation of the mean is therefore $5/\sqrt{19}$ or 1.10. And thus we are able to say with 95% confidence that the true mean efficiency will lie between 85 - 2 * 1.10 and 85 + 2 * 1.10, or the true mean will lie between 82.8 and 87.2. Of course, these formulas may be turned around to calculate the number of samples necessary to define the mean within a certain range. Thus

$$n = \frac{\sigma^2}{(\sigma_\sigma)^2}$$

Given an allowable standard deviation of the mean and some knowledge of σ, the number of necessary samples may be determined from the above equation.

REFERENCES

1. American Public Health Assoc., American Water Works Assoc., and Water Pollution Control Federation. *Standard Methods for the Examination of Water and Wastewater Including Bottom Sediments and Sludges* (New York: American Public Health Assoc., Inc. 1960).
2. Anderson, R.L., and T.A. Bancroft. *Statistical Theory in Research* (New York: McGraw-Hill Book Company, Inc., 1952).

3. Bailey, Norman T.J. *Statistical Methods in Biology* (London: The English Universities Press Ltd., 1959).

4. Bennett, Carl A., and Norman L. Franklin. *Statistical Analysis in Chemistry and the Chemical Industry* (New York: John Wiley and Sons, Inc., 1954).

5. Brownlee, K.A. *Statistical Theory and Methodology in Science and Engineering* (New York: John Wiley and Sons, Inc., 1960).

6. International Business Machines. *Proceedings of the IBM Scientific Computing Symposium on Water and Air Resource Management* (1968).

7. Johnson, Norman L., and Fred C. Leone. *Statistics and Experimental Design in Engineering and the Physical Sciences*, Vol. I and II (New York: John Wiley and Sons, Inc., 1964).

8. U.S. Department of Health, Education and Welfare. *Symposium Environmental Measurements Valid Data and Logical Interpretation*, Public Health Service Publication No. 999-AP-15 (1964).

Index

absolute density 206
absorbance 66
absorbed, light 210
absorption
  spectra, ultraviolet 89
  spectrophtometry
    atomic 42,48,65,75,77,237
    infrared 80
    molecular 65,234
    ultraviolet 88
absorptivity, molar 66,70
accuracy 17,23
acid
  and digester gas
    analysis, volatile 222
  volatile 132
acidity, alkalinity and 15
acids, fatty 222
acquisition, data 120,181
activated sludge 128,131
activation analysis,
  neutron 241
active mass 131
activity 18,124
  biological 131
  coefficient 19
  coefficient, single
    ion 232
adenosine triphosphate 131

admittance, high
  frequency 108
adsorption, carbon 35
aerial photography 172
agent
  chelating 42
  complexing 125
airborne
  fluorometric
    measurements 171
  system 166
alcohol-extract, carbon- 35
alkaline earth elements 239
alkalinity and acidity 15
aluminum 72
ammonia 16,154,255
ammonium pyrrolidine
  dithiocarbamate 42
amperometric titrimetry 105
amplifiers, operational 105
anaerobic digestion 128,132
analog transmission 120
analyses
  automated chemical 135
  continuous 118
  in situ 17
  Kjeldahl nitrogen 141
  other 160

analysis
  of data   299
  electrochemical   91
  gas chromatographic   41
  methods of   16
  parameters for   11
  pesticide   218
  phenol   219,222
  neutron activation   241
  nonspecific   215
  standard methods for   17
  temperature   182
  volatile acid and
    digester gas   222
analyzer
  galvanic cell
    oxygen   20,280
  total carbon   128,216
anions, inorganic   251
anodic stripping
  voltammetry   99,244,305
apparatus, lyophilizer   38
atomic
  absorption   42,48,65,75,
    spectrophotometry   77,237
  fluorescence
    spectrophotometry   65,77
attenuated total
  reflectance   88
AutoAnalyzer   135
  technicon   117
automated
  chemical analyses   135
  equipment   121
automation   136

Barringer Correlation
  spectrometer   177,180
benzoapyrene   37
beryllium   72
bimetallic strip
  sensors   207
biochemical oxygen   15,16,
  demand(BOD)   217
biological
  activity   131
  processes   127

bismuth   93
bolometer detectors   83
bridges, conductivity   108
Brookfield viscometer   206

cadmium   93
calcium   46
canal viscometer   206
capacitance current,
  non-faradaic   101
capture detection,
  electron   57
carbon
  adsorption   35
  alcohol-extract   35
  analyzer, total   128,216
  chloroform-extract   35
  dioxide, dissolved   115,128
  total   15,16
carrier gas   55
cathode-ray
  polarography   96
cell
  oxygen analyzer,
    galvanic   20,280
  oxygen-sensitive
    galvanic   269
centrifugation   59
characteristics
  physical   205
  primary sensor   21
  secondary sensor   24
chelating agent   42
chelation   41
chemical
  analyses, automated   135
  oxygen demand (COD)   15,16,
      141,146,156,215
  parameters   15
  potential   18
  spectrophotometry   63
chloride   46,115,155
chlorinated hydrocarbon
  pesticides   41
chlorine
  demand   15
  residual   132

chloroform-extract,
  carbon- 35
chromate 46
chromatogram 52
chromatographic
  analysis, gas 41
  columns 52
  method (oxygen), gas 273
chromatography 50
  elution 46
  gas-liquid 51,52
  gas-solid 51
  ion exchange 50,51
  paper 51
  thin layer 51,58
chromium(III) 46
chronoamperometry 96,105
chronopotentiometry 96,105
coefficient
  activity 19
  partition 39,51
  selectivity 22,44
  single ion activity 232
  thermal expansion 207
collection, sample 29
color 15
column chromatography 51
compensation,
  temperature 289
complexing agents 125
compounds
  lignin sulfonate 72
  nitrogen 255
  phenolic 91
  phosphorus 258
computer 143
  program 143
  technique 200
concentration
  factor 40
  freeze 37
conductance
  electrical 15,108,205,208
  electrolytic 107
  ionic 107

conductivity 24,107,115,209
  bridges 108
  detection, thermal 56
  electrical 208
  method, thermal 271
conductometric
  methods (oxygen) 276
  titrations 108
  titrimetry 105,108
constant
  current coulometer 266
  diffusion 100
contact electrodes 108
continuous
  analyses 118
  countercurrent extractor 41
  digestor 141
  monitoring 118
contour, thermal 182
controlled potential
  electrolysis 105
copper 46,72,93
co-precipitation 59
correlation spectrometer,
  Barringer 177,180
couette viscometer 206
coulometer, constant-
  current 266
coulometric
  analysis 105
  titration 267
  titration procedure 265
countercurrent extractor,
  continuous 41
current
  faradaic 92
  non-faradaic
    capacitance 101
  scan voltammetry 105
  voltage curve 99
curve
  current voltage 99
  fitting 300
cyanide 16

data
  acquisition 120,181
  analysis of 299

DDT 37
"dead-stop" end
  point methods 264
debubbler 154
dehydrogenase enzyme 131
demand, nutrient 16
density 11,15,205
  absolute 206
derivative
  polarogram 95
  polarography 94,95
  pulse polarography 95
detection
  electron capture 57
  limit of 21
  microcoulometric 57
  thermal conductivity 56
detectors 55
  bolometer 83
  photoconducting 176
  photovoltaic 176
  thermocouple 83
detergents 16
dialysis 38
diamagnetic 269
diethyldithiocarbamate 42
differential
  polarography 94
diffractor 169
diffusion constant 100
digester
  gas analysis, volatile
    acid and 222
  gas, sludge 222
digestion,
  anaerobic 128,132
digestor, continuous 141
digital transmission 120
dioxide, dissolved
  carbon 115,128
discriminator, 173,177,
  Fraunhofer line 179
dispersion 81
dissolved
  carbon dioxide 115,128
  gases 261
  oxygen 16,115
  solids 205

distillation 59
  vacuum 38
dithiocarbamate,
  ammonium pyrrolide 42
dithizone 66
dropping mercury
  electrode 93,278

effect
  photoconduction 176
  photoelectromagnetic 176
  photoemission 176
  photovoltaic 176
efficiency factor,
  quantum 72
effluent
  standards 8,9
  sulfite 72
effluents, industrial 191
electrical
  conductance 15,108,205,208
  conductivity 208
  thermometer 207,208
electrode
  oxygen membrane 284,289
  system, potentiometric
    membrane 106
  systems, oxygen
    membrane 279
  voltammetric membrane 279
electrodes
  hanging-drop mercury 101
  ion-selective 241,253
  potentiometric membrane 254
  solid-state membrane 106
electron capture
  detection 57
electrochemical analysis 91
electrolysis, controlled
  potential 105
electrolytic conductance 107
electrophoresis 60
elements, alkaline earth 239
elution chromatography 46
emission
  infrared 176
  spectroscopy 240

end point methods,
"dead-stop" 264
energy, molar free 18
enzyme, dehydrogenase 131
equation, Ilkovic 92
equipment, automated 121
evaporation 59
exchange
 chromatography, ion 50,51
 ion 43
excitation, selective 171
expansion coefficient,
 thermal 207
extensive 18
 measurements 18
 properties 19
extract, carbon-
 alcohol- 35
 chloroform 35
extraction,
 liquid-liquid 38,140
extractor, continuous
 countercurrent 41

Fabry-Perot filter 180
factor
 concentration 40
 quantum efficiency 72
faradaic current 92
fatty acids 222
filter, Fabry-Perot 180
filterable residues 15
flame
 photometry 77
 techniques 239
flow systems 150
fluorescence 71,171
 radiation 70
 spectrophotometry 70
  atomic 65,77
  molecular 65,70,72
fluorescent
 substances 173
fluoride 49,251
fluorometric
 measurements,
  airborne 171
 reagents 71

formation, ion pair 233
Fortran IV 143
Fraunhofer line 189
 discriminator 173,177,179
free energy, molar 18
freeze concentration 37
frequency admittance,
 high 108

galvanic
 cell oxygen analyzer 20,280
 cell, oxygen-sensitive 269
 membrane electrodes 281
galvanostat 103
gamma radiation 131
gas
 analysis, volatile acid
  and digester 222
 carrier 55
 chromatographic analysis 41
 chromatographic method
  (oxygen) 273
 exchange separation 261
 exchange unit 267
 liquid chromatography 51,52
 sludge digester 222
 solid chromatography 51
gases, dissolved 261
glass electrodes 105
grating 81
gravity, specific 206
grease and oil 16

half-wave potential 93
hanging-drop mercury
 electrodes 101
hardness 15
heavy metals 16
Henry's Law 262
herbicide 49
high frequency
 admittance 108
hydrocarbon
 pesticides, chlorinated 41
 polynuclear 37

identification
 of pesticides 226
 of phenols 226

Ilkovic equation 92
imagery, multispectral 191
industrial effluents 191
infrared
 absorption
  spectrophotometry 80
 emission 176
 near 83
 radiation 138
 sensing 176
 spectra 83
inorganic anions 251
in situ analyses 17
intensive 18
 properties 19
interferences 17
ion
 activity coefficient,
  single 232
 exchange 43
 exchange
  chromatography 50,51
 pair formation 233
 selective
  electrodes 241,253
ionic conductance 107
iron 72,156
iron(III) 46,48
irreversible process 99
isobutyl ketone, methyl 42

Kjeldahl nitrogen 158
 analyses 141

laser 166
 neon-pulsed 173
 scanner system 170,171
lead 93
lead(II) 48
light
 absorbed 210
 scattered 210
lignin sulfonate
 compounds 72
limit of detection 21
line discriminator,
 Fraunhofer 173,177,179

linear
 potential sweep
  stripping 245
  programming 307
liquid-liquid
 extraction 38,140
long term stability 23
lyphilization 38
lyophilizer apparatus 38

magnesium 46
manganese 158
manual procedures 145
mass, active 131
material, organic 36
matter
 particulate 205
 suspended 15
 volatile 205
measurement
 remote 167
 turbidity 170
measurements
 airborne fluorometric 171
 extensive 18
membrane electrode
 system, potentiometric 106
 systems, oxygen 279
 voltammetric 279
membrane electrodes
 galvanic 281
 oxygen 284,289
 potentiometric 254
 solid-state 106
membrane processes 59
membranes, plastic 294
mercury electrode
 dropping 93,278
 hanging-drop 101
metal
 heavy 16
 pollutants 231
metaphosphate 46,258
method
 paramagnetic 269
 thermal conductivity 271

method (oxygen)
  gas chromatographic   273
  radiometric   275
methods
  "dead-stop" end point   264
  for analysis, standard   17
  of analysis   16
methods (oxygen)
  conductometric   276
  voltammetric   278
methyl isobutyl ketone   42
microbial oxidation   215
microcoulometric
  detection   57
molar
  absorptivity   66,70
  free energy   18
molecular
  absorption
    spectrophotometry   65,234
  fluorescence
    spectrophotometry   65,70,72
  sieves   51
monitoring
  continuous   118
  systems   115
monochromator   81
multispectral
  imagery   191
  scanner   181,191
  sensor   177
  sensor system   177
municipal
  sewers   9
  wastes   10

near infrared   83
neon-pulsed laser   173
Nernst
  equation   105
  partition law   39
neutron activation
  analysis   241
nickel(II)   48
nitrate   16,157,255
nitrite   16,157,255

nitrogen   16
  analyses, Kjeldahl   141
  compounds   255
  Kjeldahl   158
  organic   16
non-faradaic
  capacitance current   101
nonspecific
  analysis   215
  parameters   15
  tests   11
nuisances   16
number of samples   28
nutrient
  demand   16
  specific   16

objectives   7
odor   11,15,35
oil   35
  grease and   16
  leak   190
  pollution   190
  slicks   190
operational amplifiers   105
optical
  mechanical scanner   177
  sensing techniques,
    remote   165
  turbidity   212
optimization   127
organic
  material   36
  nitrogen   16
  phosphorus   16,141
  pollutants   215,218
  orthophosphate   16,46,258
oscillographic
  polarography   96,99
other analyses   160
oxidation
  microbial   215
  reduction potential   115
oxygen
  analyzer, galvanic
    cell   20,280
  demand (BOD),
    biochemical   15,16,217

327

oxygen (cont)
demand (COD),   15,16,141,
  chemical   146,156,215
  dissolved   16,115
  membrane electrode 284,289
  systems   279
  sensitive galvanic
  cell   269

pair formation, ion   233
paper chromatography   51
  chromatography   51
  probability   309
paramagnetic   269
  method   269
parameters
  chemical   15
  for analysis   11
  non-specific   15
  physical   11,15
  physiological   11,15
particulate matter   205
partition coefficient 39,51
pesticides   16
  analysis   218
  chlorinated hydrocarbon 41
  identification of   226
pH   115
phenol   16,35
  analysis   219,222
phenolic compounds   91
phenols, identification
  of   226
phosphate   158
phosphorus   16
  compounds   258
  organic   16,141
photoconducting
  detectors   176
photoconduction effect   176
photoelectromagnetic
  effect   176
photoemission effect   176
photography, aerial   172
photometry, flame   77
photovoltaic
  detector   176
  effect   176

physical
  characteristics   205
  parameter   11,15
physiological
  parameters   11,15
  property   11
plastic membranes   294
plate theory   52
plume, thermal   182
polarogram, derivative   95
polarography   92,243
  cathode-ray   96
  derivative   94,95
  pulse   95
  differential   94
  oscillographic   96,99
  pulse   95
pollutants
  metal   231
  organic   215,218
pollution
  oil   190
  thermal   187
polyethylene   285
polynuclear
  hydrocarbons   37
polyphosphate   16,258
polypropylene   285
potential
  chemical   18
  electrolysis,
  controlled   105
  half-wave   93
  oxidation-reduction   115
  sweep stripping, linear 245
potentiometric
  membrane electrode   254
  system   106
  titrimetry   105
potentiostat   103
precipitation   59
precision   24
preservation, sample   30
primary sensor
  characteristics   21
prism   81
probability paper   309

procedure
  coulometric titration 265
  manual 145
process
  irreversible 99
  reversible 96
processes
  biological 127
  membrane 59
programming, linear 307
programs, sampling 27
properties
  extensive 19
  intensive 19
property, physiological 11
proportioning pump 137
pulse polarography 95
  derivative 95
pump, proportioning 137
pyrometers, radiation 207
pyrroldine
  dithiocarbamate,
    ammonium 42

quantum efficiency
  factor 72

radiation
  fluorescence 70
  gamma 131
  infrared 131
  pyrometers 207
  solar 115
radiometric method
  (oxygen) 275
random sampling 27
reagents, fluorometric 71
reflectance, attenuated
  total 88
reflector 169
remote
  measurement 167
  optical sensing
    techniques 165
  sensing 165,200
residual chlorine 132
residues, filterable 15

resistance thermometers 121
resolvability 94
response time 21
results, statistical 318
reversible process 96
rubber, synthetic 285

salinity 15,107,108
salting
  in 20
  out 20
sample
  collection 29
  preservation 30
samples, number of 28
sampling
  programs 27
  random 27
  sites 28
scanner
  multispectral 181,191
  optical-mechanical 177
scattered, light 210
scattering 167
secondary sensor
  characteristics 24
selective excitation 171
selectivity 22,77,106
  coefficient 22,44
self-purification 9
semiconductor 123
sensing
  infrared 176
  remote 165,200
  techniques, remote
    optical 165
sensitivity 17,21,77
sensor
  characteristics,
    primary 21
  characteristics,
    secondary 24
  multispectral 177
  system, multispectral 177
sensors 121
  bimetallic strip 207
separability 94

separation

  gas-exchange 261

  solubility 36

  techniques 33

sewers, municipal 9

sieves, molecular 51

silica 160

silicate 46

single ion activity

  coefficient 232

sites, sampling 28

sludge

  activated 128,131

  digester gas 222

solar radiation 115

solid-state membrane

  electrodes 106

solids

  dissolved 205

  suspended 128

solubility separation 36

specific

  gravity 206

  ion electrode 106,124

  nutrients 16

specificity 75

spectra

  infrared 83

  ultraviolet absorption 89

spectrometer, Barringer

  correlation 177,180

spectrophotometer 66

spectrophotometry 42,48,65,

  atomic absorption 75,77,237

  atomic fluorescence 65,77

  chemical 63

  infrared absorption 80

  molecular absorption 65,234

  molecular

    fluorescence 65,70,72

  ultraviolet

    absorption 88

spectroscopy

  emission 240

  fluorescence 70

stability, long term 23

standard methods for

  analysis 17

standards

  effluent 8,9

  stream 8

statistical

  results 318

  testing 316

  tests 315

stream standards 8

strip sensors,

  bimetallic 207

stripping 99

  linear potential sweep 245

  voltammetry 101

  anodic 99,244,305

sublimation 59

substances, fluorescent 173

sulfate 46,251

sulfide 16,124,251

sulfite 16,251

  effluent 72

sulfonate compounds,

  lignin 72

surface tension 205

suspended

  matter 15

  solids 128

sweep stripping, linear

  potential 245

synthetic rubber 285

system

  airborne 166

  laser-scanner 170,171

  multispectral sensor 177

  potentiometric membrane

    electrode 106

  three-electrode 102

  two-electrode 101

systems

  flow 150

  monitoring 115

  oxygen membrane

    electrode 279

taste 15,35

technicon AutoAnalyzer 117

techniques

  remote optical sensing 165

  separation 35

teflon 285
telemetering 120
telemetry 121
temperature 24,115,205,206
 analysis 182
 compensation 289
tension, surface 205
test, Winkler 20,264
testing, statistical 316
tests, statistical 315
theory, plate 52
thermal
 conductivity detection 56
 conductivity method 271
 contour 182
 expansion coefficient 207
 plume 182
 pollution 187
thermistor 121,207,208
thermocouple 121,208
 detectors 83
thermometer
 electrical 207,208
 resistance 121
thin-layer
 chromatography 51,58
thiosulfate 251
three-electrode system 102
titration
 coulometric 267
 procedure, coulometric 265
titrations,
 conductometric 108
titrimetry
 amperometric 105
 conductometric 105,108
 potentiometric 105
total carbon 15,16
 analyzers 128,216
total reflectance,
 attenuated 88
toxicity 16
transducers 121
transmission
 analog 120
 digital 120
triphosphate, adenosine 131
t-test 315

turbidity 15,115,205,210
 measurement 170
 optical 212
two-electrode system 101
ultraviolet absorption
 spectra 89
 spectrophotometry 88
unit, gas-exchange 267

vacuum distillation 38
viscometer
 Brookfield 206
 canal 206
 couette 206
viscosity 205,206
volatile acid 132
 and digester gas
  analysis 222
volatile matter 205
voltage
 curve, current 99
 scan voltammetry 105
voltammetric
 membrane electrode 279
 methods (xoygen) 278
voltammetry
 anodic stripping 99,244,305
 current-scan 105
 stripping 101
 voltage-scan 105

wastes, municipal 10
water quality
 Act 8
 criteria 8
Wheatstone Bridge 108
Winkler test 20,264

zeolites 43
zinc 93